IN SIGHT

IN SIGHT

My Life in Science and Biotech

Foreword by
Molly Shoichet

JULIA LEVY

UNIVERSITY OF TORONTO PRESS
Toronto Buffalo London

Rotman-UTP Publishing
An imprint of University of Toronto Press
Toronto Buffalo London
utorontopress.com

Library and Archives Canada Cataloguing in Publication

Title: In sight : my life in science and biotech / Julia Levy.
Names: Levy, Julia G., 1934– author.
Description: Includes bibliographical references and index.
Identifiers: Canadiana (print) 2020029895X | Canadiana (ebook) 2020029900X |
    ISBN 9781487508319 (cloth) | ISBN 9781487537999 (EPUB) |
    ISBN 9781487537982 (PDF)
Subjects: LCSH: Levy, Julia G., 1934– | LCSH: Microbiologists – Canada –
    Biography. | LCSH: College teachers – Canada – Biography. | LCSH: Chief
    executive officers – Canada – Biography. | LCGFT: Autobiographies.
Classification: LCC QR31.L48 A3 2020 | DDC 579.092–dc23

ISBN 978-1-4875-0831-9 (cloth)   ISBN 978-1-4875-3799-9 (EPUB)
                                     ISBN 978-1-4875-3798-2 (PDF)

Printed in Canada

We acknowledge the financial support of the Government of Canada, the
Canada Council for the Arts, and the Ontario Arts Council, an agency of the
Government of Ontario, for our publishing activities.

Canada Council   Conseil des Arts
for the Arts   du Canada

ONTARIO ARTS COUNCIL
CONSEIL DES ARTS DE L'ONTARIO
an Ontario government agency
un organisme du gouvernement de l'Ontario

Funded by the   Financé par le
Government   gouvernement
of Canada   du Canada

MIX
Paper from
responsible sources
FSC® C016245

To Ed, my life partner

Contents

Foreword

Dr. Julia Levy is a remarkable woman with an enviable story, but one that includes great loss. She was a trailblazer in science and the biotechnology industry, seamlessly transitioning between academia and industry. Her story would be just as fascinating if it were happening today, and is even more engaging in that it took shape in the 1960s, at a time when women typically stayed home. It certainly wasn't a time when women pursued PhDs in science, went on to be university professors, and then became CEOs of billion-dollar companies.

Dr. Levy immigrated to Canada from Jakarta after the Dutch surrendered to the Germans. Her British mother gave up her citizenship when she married her Dutch father – the norms of the 1940s. Politics and the war provide the backdrop for Dr. Levy's immigration to Vancouver, Canada. She portrays herself as different and curious, often getting into trouble as a child. She grew up in a time when most girls were deterred from studying science in university, and she succeeded despite the stereotypes. Her story reflects how people in that era did not typically dwell on the bad, but rather figured out how to make life work.

This very practical approach to life resonates throughout Dr. Levy's autobiography. Scientific advances are woven into her own life story. For example, as an undergraduate student, Julia was intrigued with virology and bacteriology after learning how William Jenner made an observation that led to vaccines. Moreover, Julia enjoyed a wonderfully fascinating environment during her PhD and post-doctoral training, as she was surrounded by Nobel laureates.

Dr. Levy's biography is honest and reflects a woman with integrity and a deep dedication to science and its application to products that impacted patients' lives significantly. Dr. Levy beat all of the odds and was definitely a trailblazer in Canada, North America, and indeed the world. Dr. Levy had a junior faculty position at the University of British Columbia at age twenty-four, already with a baby in tow. This was a remarkable achievement then and would be equally remarkable today.

She developed an idea, brought it to fruition, and built a billion-dollar company along the way.

Dr. Levy emphasizes the importance of a deep scientific grounding as the backbone for the products that she brought to market. Having deep knowledge and understanding of her product made the difference between Dr. Levy's success and her competitors' failures. These ideas resonate with me and my own pursuits to bring products to market. The innovation economy, on which we all base our future, is built on these ideas modelled by Dr. Levy – and at a time before "innovation economy" was a buzzword.

She describes her life leading up to the founding of QLT in a matter-of-fact tone. Her writing is open and easy to read. We learn about the many trials and tribulations of bringing a blockbuster drug to market. We learn about all the landmines along this journey and we discover that even after success landmines continue to emerge. I enjoyed reading this book once and know that I will return to it repeatedly to get inspiration and guidance as I pursue my own journey to translate ideas into products.

Dr. Levy had the tenacity to succeed in a man's world and the wherewithal to forge a new way of advancing knowledge from the lab to a spin-off company that was one of Canada's great successes of the 1990s and 2000s.

I am grateful to Dr. Julia Levy – she cultivated an environment that made it easier for me and my generation to succeed in science. Society has certainly advanced since she forged new ground, yet women still face an undercurrent of discrimination in academia and industry in terms of promotions and leadership opportunities. I have benefited from Dr. Levy's wisdom. I've learned more about science, entrepreneurship, and life and feel richer for having read her autobiography.

Molly Shoichet, PhD, O.C., O.Ont., FRS
Professor, University of Toronto

IN SIGHT

CHAPTER ONE

Waiting for the Other Shoe to Drop

News that Holland had surrendered to the invading Nazis reached the Dutch East Indies on May 15, 1940. That date was also my sixth birthday.

On May 10, panzers overran the largely undefended border between neutral Holland and Germany while the Luftwaffe flattened the city of Rotterdam. The country surrendered on May 14. My family at that time was living in Batavia (now Jakarta) on the island of Java (now in Indonesia), which was part of the Dutch East Indies.

I remember the day well – my expectations were high for the promised party with cake and candles after my sister Pamela returned from school.

But Pamela didn't go to school that day, and my mother was clearly distracted. She spent a lot of time in conversation with the lady who lived next door. And the servants were nowhere to be seen, keeping out of sight in their quarters.

We lived in a comfortable bungalow with large covered verandahs front and back in a middle-class neighborhood. The house servants – our *amah*, a cook, and two maids – lived in a separate building at the rear of our back garden, connected to our house with a covered walk.

I knew there was something wrong, and it made me nervous.

My mother told me a few years later that the white population in Indonesia was terrified that day after receiving warnings on radio that the Javanese might well rise up against their Dutch masters now that Holland had been defeated. Women were warned to keep their children close and advised to carry bags of pepper around with them in case they were accosted.

She also told me what a pest I was that day. I had expected to be the center of attention, and I wasn't. Worse, the number of children coming to my party had dwindled. I misbehaved terribly about the cake, insisting that all six candles on it were mine, along with the chocolate roses that served as holders for the candles. Having acquired the candles I then proceeded to eat one before my mother stopped me.

My father, Guillaume Albert Coppens, was Dutch, and my mother English. He worked in a Dutch merchant bank that had a significant presence in the Dutch East Indies, as well as other countries in the Far East. My sister was born in Sumatra, and I was born in Singapore two years later.

We moved to Java when I was four. Many of my memories of that time involve animals. We had a mynah bird named Pippy that sat on my mother's sandaled foot and talked as well as any parrot. A friend of my mother's had a large (to me) tortoise that ate lettuce from our hands. We had a much-loved dachshund named Bettica.

Snakes were bad news in Java. I remember two snake experiences. One was at a hill station when we were on a walk. The snake didn't threaten us, but we were terrified and ran away. Another time one of the servants found a large snake in the back garden, curled under a banana tree. They killed it, but I remember seeing its long black and white body laid out on the grass. I had nightmares for a while after that, waking to see the dark wrinkles in the sheet covering me and thinking they were snakes.

Another traumatic animal experience I recall was one for which I got into trouble. I remember waking early from my afternoon nap and wandering out onto our back verandah. My sister Pamela wasn't there; she was probably at school. I saw our cook and another of the Indonesian staff holding two live chickens. I suppose they were slaughtering them for our evening meal. I'm not sure what they did, they probably choked and then beheaded them, but they placed a couple of bricks on the chickens' necks and left them on the ground. The birds were fluttering in their death throes. The cook and his friend disappeared back into their quarters and I was left watching the birds' flailing spasms. I was aghast at the cruelty I was seeing and, without thinking, I ran across the lawn and lifted the bricks from their necks. Although the birds were dead they were far from immobile, and I remember a storm of white wings and feathers and blood around me. Then the cook came back, shouting at me and grabbing the birds. Then my amah hurried up, took me in, and gave me a bath because I was splashed with blood.

Later my mother scolded me severely for interfering and invading the servants' quarters, which were off bounds for us.

On another occasion I was playing tug-of-war with a friendly dog when it accidently bit me on the finger. Although it was a tiny puncture wound, it created panic. Everyone who lived in the tropics then was extremely fearful of diseases like tetanus or rabies. A bite of that kind could carry either of these. I was taken to a doctor and threw a tantrum, refusing to remove my underpants for him to check me over, lying on the examining table clinging to my skimpy little pants and screaming that I had a bite on my finger, why should I take off my clothes.

My father took to calling me "young Hitler" jokingly at that time (this was made in the 1930s, when most of the world regarded Hitler as an obnoxious flash in the pan). Since I didn't know who Hitler was, I was not offended.

My mother tried to enter me in a kindergarten the year I was five. My sister was in school then and I was avid to learn everything she was taught. She was a willing teacher. I don't believe I thought kindergarten was beneath me but I remember not being enthusiastic. My mother dropped me at the place. I remember it clearly even though I was there only once. It was a regular house, not a school. There were (to me) high stairs leading up to the main floor, where the classroom was. There were small desks and perhaps a dozen and a half children, boys and girls, all of them Dutch. There were two middle-aged women in charge.

I got into an incident that first morning with a little boy who was sitting beside me. He came over to my desk and took a lump of plasticene off my table. I went over to his table and grabbed it back. One of the ladies saw me do it and came over and scolded me in Dutch. I tried to explain what had happened but she wouldn't listen.

At some point we were taken out into the front yard to play. I was sulky and full of righteous indignation. There were pillars under the stairs leading to the front door and I went behind one of these pillars and hid. The other children were called in and I sat tight. I'm not sure how long I sat there, but I remember starting to get really worried because I didn't know quite what to do. I had kind of burned my bridges and I had no idea how to get home. Finally I saw my mother arriving in one of the horse-drawn cabs often used instead of taxis. I ran out. I think she was so relieved to see me that I escaped serious punishment. But I never went back to kindergarten. I suppose I was expelled. Pamela tutored me when she got home from school and, I think, prepared me better for school than any play school would have done.

I don't remember my sister getting into trouble as much as I did, but that can probably be attributed to my lack of concern for her problems. I do remember that she got her hands on a pair of scissors at one point and cut my hair. She got in trouble as a result. I cut the hair on my teddy bear. I was not in trouble for that, but his hair didn't grow back like mine did and that upset me.

When I think of those months after my sixth birthday and try to remember details and thoughts around that period, I don't remember anything out of the ordinary. But I know there was tremendous fear and uncertainty about the future for Dutch nationals in Indonesia in the months following my sixth birthday. I know from talking to my mother that the time leading up to our departure for Canada was fraught with fear. The European war was raging, Britain was being blitzed, and large parts of Europe were occupied by the Germans. And the Pacific arena was warming up. Japanese troops were already in China, and everyone living in the east knew that there would soon be war in the Pacific too.

My father was in the Dutch reserve army, as were all able-bodied Dutchmen in the colonies, so he was duty-bound to remain in Indonesia. The Dutch government in exile was in Ottawa. My mother had given up her British nationality when she married my father, as was the law in those days. Australia and Canada offered the two most obvious places for us to emigrate to if we were to escape the war.

My mother had an aunt who had married and emigrated with her husband to Saskatchewan, where they started a dairy farm. The depression had ruined them, and they had come to Vancouver to try to survive there. These were the only kin either my mother or father had who were not in the theater of war. So my parents decided that we should find a way to get to Vancouver. My father would remain in Batavia.

Certainly, all the planning and discussion that went on between my parents about the future was not shared with either my sister or me. I became aware that we were leaving the day my father came home with the tickets and told us we had eight days to pack our things and be ready to leave. He had booked us on a Dutch passenger liner that was heading to Portland. From there we would take a train to Vancouver. The trip would take about five weeks, with stops in Borneo, Manila, and Hawaii.

We brought very little with us. I was allowed only two toys, and I chose my teddy bear and my Walt Disney Snow White dwarf, Happy. The hairless teddy bear was my most cherished toy, but I'm not sure why I picked Happy. Pamela had received a Walt Disney doll as well, Dopey, and I'm not sure why she chose him, either. Her other choice was an appalling doll with a black face, then called a gollywog – a more politically incorrect toy thing I can't imagine, but she loved her golly. I had received a Raggedy Ann when Pamela received her golly. I left Raggedy Ann in Indonesia, although I did love the bloomers she had under her floppy dress, because they came off and I was always keen to denude any doll I received.

On the ship my mother always carried an oilskin package around her waist containing money and jewelry. She told us that she carried the money in case we were torpedoed and ended up in the water. The money was from her dowry, something in the range of $10,000. In today's money, this would be equivalent to about $150,000. So we weren't poor.

Although the war was not yet active in the Pacific, news of German U-boats was everywhere and the ship was in blackout at night. We had lifeboat drills regularly to remind us of the war. I remember feeling a little excited rather than scared about the possibility of U-boat attacks.

My mother told me later that she had become friendly with the quartermaster on the boat, and that he had told her the boat had actually been set on fire by a fifth columnist on board. The man had been caught and put into the prison cell on the ship until we reached Hawaii, and then handed over to the police.

We arrived in Portland towards the end of August and took an overnight train to Vancouver. My mother, thinking Canada was a country of ice and snow, had prepared us well. In the week before we left Indonesia she rushed around and had special clothing made for Pamela and me: heavy serge skirts, knee socks (I don't know how she ever found them), and woolen jackets. Undeterred by the heat in our little Pullman berth, she made us dress in these clothes the morning of our arrival. I sweated, and looked out the window as the train snaked through the town of White Rock, south of Vancouver, and I saw children in bathing suits running around on the beach and swimming.

We arrived at the CNR station on Main Street and were met by my mother's uncle, Uncle Dick. He was a tall, thin, elegant man in his late sixties with a trim white mustache. My mother's aunt, Jess, had died a few weeks earlier. We must have left our luggage at the station because we walked from there, through the downtown area and into the West End, where we went from door to door looking for a rooming house that would take children. I felt hotter than I ever had in Java. Pamela and I trailed after my mother, wilting.

We were rejected by many rooming houses, but finally a kindly lady took pity on us and rented us an attic room with two single beds in it. Pamela and I shared one bed.

The rooming-house business was booming in 1940. The war effort was gearing up. Men were pouring into Vancouver to get jobs in the burgeoning shipyards, glad for employment after the lean years of the depression.

We only stayed a few days in the rooming house. My mother found a basement apartment in an old house on Comox Street. That house still stands today and is close to where I live now. She enrolled us in a public school, Lord Roberts, two blocks from our apartment. Pamela and I lasted one day there. My mother, being British, was not happy with the way the teachers spoke or the rough way the children played, and she took us out of school. I was just as glad because I had already found myself in trouble.

We had been given pencils. Like many of my fellow students I immediately started moving my pencil back and forth in the indentation that ran along the top of my desk, presumably put there to prevent pencils from sliding off the slightly slanted desk top. A background noise must have arisen from a dozen or so busy fingers moving our pencils back and forth. The teacher shouted at us to stop, and said what sounded to me like "raddle," to "stop raddling our pencils." I didn't know what raddle meant – I was not familiar with the North American accent and was used to speaking Dutch as well as English. Everyone else stopped. I did not.

Then the teacher was standing over me as I sat there, looking up at her in all innocence, still "raddling" my pencil. She took my pencil away and I was shamed and heartbroken.

My mother found a peculiar marginal girl's school in the West End called Saint Marina. It was on its last legs. The headmistress, Miss Seymour, became senile the third year we were there and the school closed after she started appearing in class in her nightgown. The school was in a big old late-Victorian house built on a fairly grand scale. There were only four classrooms, with grades one to six in two rooms, divided by a sliding door. Higher grades were in classrooms on the second floor, and I never saw them. The grades four to six classroom also served as the dining room, so the students sat around a wide long table on which we ate our lunch and dinner after classes.

The most traumatic episode I experienced that first year was my sister's fault. The lunchroom at school was crowded and the principal had put in a small table and chair for Pamela and me beside hers at the head of the big dining table. On the day in question I had heard someone call another person a "sweetie-pie." I had been charmed by the sound and at lunch that day I leaned towards Pamela and said, "You're a sweetie-pie."

Pamela's eyes grew large and she put her hand to her mouth. "That's a terrible word," she said, almost whispering. "You could get expelled for saying things like that."

I was stunned. To make matters worse, I had noticed Miss Seymour, the headmistress, glancing down at us just as I spoke. I knew she had heard what I'd said.

I spent the next few days waiting for the other shoe to drop. I moped around. My mother noticed and probed me gently about what was troubling me. I couldn't speak of it. Finally, one night, I was standing on top of the closed toilet seat while my mother dried me after my bath and I cracked. I asked if it was true that one could get expelled for saying bad words.

My mother related later that her thoughts were, "Oh dear, what has she done now?" She responded very evenly that she didn't think it likely, that one could explain such a lapse if one was sorry. Then she asked me what I'd said.

I was so ashamed I couldn't bring myself to say sweetie-pie again so I just shook my head. But she pressed me.

I said the words so fast she thought it sounded like "swishpa" or something like that. She thought, "oh dear, it's some Canadian word I've never heard before." She finally got it out of me and had trouble keeping a straight face. But she warned me that if I ever heard a word I didn't understand it would be better if I passed it by her first. Good advice. The next bad word I heard was "shit," from a little boy who lived across the street. I knew it was a bad word because of the way he used it, but I thought he'd said "sheep" and was really confused.

My first teacher was a young woman with a gentle face and manner called Miss Burton. I liked her very much and excelled. I was moved to grade three at the end of the first year. My sister had taught me well. She was in grade four and moved to the other classroom. Miss Burton joined the Air Force through the WAF program and came to say goodbye to the class in her uniform. We were all very proud of her.

At the end of that year I was promoted to grade five. My mother refused to allow this. She did not want Pamela and me to be in the same grade.

My mother had not been idle. As I recall, we did not stay long in the basement apartment, probably less than a year. But this was long enough for us to discover we had a family of pack rats sharing the basement. We would find vegetables my mother had stored in the pantry, neatly lined up in the toilet. My mother thought I was responsible, and I was offended. She apologized when we discovered it was pack rats. They were trapped and killed and I was sorry for them.

After the pack rats my mother bought a house in the suburban area of Dunbar at Thirty-Eighth and Blenheim, a tiny four-room house with living room, kitchen, and two bedrooms. It was heated with a sawdust furnace that blasted heat up through a vent in the small hallway that the two bedrooms and bathroom opened into. The basement was filled with sawdust every fall and the furnace had to be fed regularly.

I'm not sure how my mother wangled it, but she got herself a job as a masseuse in a very genteel practice for ladies with stiff backs or other afflictions and run by a large prim lady called Miss Markham, who always wore a hat. Her practice was in the old Birk's building at Granville and Georgia. My mother had been to an institute in England called the Liverpool Training College and had trained, as far as I can conclude, as a gym teacher. But she had learned some anatomy and possibly some basic massage techniques, enough to get her a job.

My mother cultivated woman friends, whether as neighbors or friends of friends. And they would help each other. I think many of the opportunities she had came through this network.

With our move to the suburbs from the West End and my mother getting a job, Pamela and I could no longer continue in the school as day girls, so my mother put us in as boarders. I hated it. The dormitory where the boarders slept was the top floor of the house: an enormous long room, the kind one sees in orphanages, rooms with a row of beds on either side. It was cold, and sometimes my feet were so cold I couldn't sleep.

We were allowed home for weekends, and I lived for those days. Pamela and I would take the streetcar from Davie and Bute, changing twice, to get to Forty-First and Blenheim and walk from there, about half a kilometer. We'd get home before my mother and stoke the furnace, Pamela on a chair pouring buckets of sawdust into the furnace and me fetching the buckets from the stored piles.

Through 1941, my father was able to continue sending money and letters. After Pearl Harbor, communication became much more difficult. He sent a beautiful carved camphor-wood chest, the kind one sees in stores selling oriental furniture and artworks. I thought it wonderful and its smell of camphor magical. It was the most beautiful thing we had in our sparsely furnished small house. We stored extra blankets or winter coats in it.

What I didn't know was that, in a covering letter, my father had cryptically told my mother (cryptically because censorship of all mail was in place) that the chest contained something of considerable value and that under no circumstances should she sell it. My mother and Uncle Dick spent hours contemplating the chest trying to figure out where some hidden compartment might be. They never did solve the riddle. My father, although imprisoned by the Japanese for two and a half years, survived the war and made it to Canada and was able to reveal the chest's secret. Under the brass lock he had hollowed out an additional space, in which he'd secreted a bag of diamonds.

Over the years, I've often thought about my mother in those first years we were in Canada. She was really on her own: a woman who had never worked or cooked a meal, nor expected to have to, with two small children in a strange country about which she knew next to nothing. After the Japanese invaded Java in 1942 my mother lost communication with my father. She didn't know if he was dead. Her future was completely uncertain. When I talked to her about that time, she was always very Britishly matter-of-fact, her attitude being "you do what you have to and don't go on about it."

We were in Canada illegally, it turned out. We came on visitors' visas that were good for six months. My mother struck up a friendship with the director of the Dutch consulate. He was sympathetic to my mother's situation and was not about to recommend her deportation; as well, there was nowhere to send her. She had to get our visas renewed every six months. This continued until after my father came back to us and we applied for Canadian citizenship.

Through those years of the Second World War, my mother provided Pamela and me with a secure childhood. Somewhere in the early forties she interviewed for a government job at the then Workman's Compensation Board, and was given a position as a physiotherapist. There must have been a terrible shortage of physiotherapists at the time. With logging and shipbuilding going strong, there were plenty of industrial accidents. Otherwise, I don't think her credentials would have been accepted. But she kept that job until she retired with a pension.

We never had any reason to feel insecure. We went to school and came home for weekends. During the summer we went through a varied group of child-minders, all emanating from my mother's network of women friends. Or we went to a summer camp at

Roberts Creek called Kewpie Camp. We managed to accumulate pets – guinea pigs, rabbits, dogs, cats. My mother fed them all during the week while we were at school.

I never questioned the fact that my mother worked, even though none of the other children on our block had mothers who did. They had fathers who went to the office. I did feel we were different, and I was all right with the difference. My mother had created a home in which I felt absolutely secure and happy. Our exclusively female existence was a very comfortable one.

CHAPTER TWO

Does Your Dog Always Sit in the Front Seat?

The Dutch East Indies was invaded by Japan on February 28, 1942. The islands formally surrendered on March 12, and all men serving in the military were imprisoned according to the Geneva Convention. There had been almost no resistance to the invasion.

My mother told us what had happened, that our father was a POW and we wouldn't hear from him for a while. I don't remember feeling an undue amount of anxiety. My father was a distant figure. He had not been part of our new lives for a year and a half. Our lives in Canada had a pattern that did not change, and he was not part of that life.

We did change schools. My mother found another girls' school located on Twenty-Seventh and Granville, just two blocks away from York House, another private school that still exists today. Our school, Queen's Hall, does not. Miss Bodie, the headmistress, owned the school and ran it with her sister, Mrs. Whitmore.

In comparison to St. Marina, Queen's Hall was a going concern. There were around thirty boarders and at least an equal number of day girls. Most of our fellow boarders were from homes in remote parts of the province, many of them the children of ranchers and fruit growers from the Okanagan Valley. Several were children of divorced parents.

The main building of the school was a converted grand Shaughnessy house with four big airy classrooms on the main floor that covered grades one through six. There was also the annex, the original coach house, separate from the main house, with two large rooms on the main and upper floor that accommodated grades seven through twelve. Shaughnessy in Vancouver is the equivalent of San Francisco's Nob Hill, and is where, in the past, the province's lumber barons built their mansions.

I did not enjoy boarding school. On my first day at Queen's Hall I had a dreadful disappointment. We were shepherded in to dinner and to my delight I saw what I thought was a plate with lettuce and some canned fruit on it and what looked to me like a big scoop of ice cream. "Ice cream as a main course," I thought, "what a treat!" Imagine my horror when I took a forkful of this delicacy only to discover cottage cheese, which I'd never seen before. I dislike cottage cheese to this day.

I developed all my dislike of certain foods at boarding school. Most of the food was ghastly; mucilaginous macaroni, baked beans, lumpy cream of wheat, and starchy puddings like tapioca (aka fish eyes in glue), cornstarch, and rice stand out in my memory, all foods I won't eat now. Leftover toast from breakfast was stored overnight in a metal container and reheated the next day, and tasted of tin. It was referred to as "bread box toast" and avoided whenever possible. After several days, leftover bread box toast was converted to bread pudding, one of the foulest desserts the cook managed to create, metallic and watery.

In spite of the terrible food, I got chubby at that school, probably because I filled up on bread, potatoes, and milk.

I was an untidy child. My knee socks were always falling down. We wore layers of underclothing, woolen underpants with heavy black bloomers on top, a woolen undershirt, then a cotton shirt and tie and a belted gym tunic over everything. These garments often bunched up if insufficient care was put into arranging them when getting dressed. I tended to get things a bit bunched up. Miss Bodie called us all by our surnames, after the British style. Pamela and I became Coppens I and Coppens II. Miss Bodie on one occasion called me out of line, because my clothes were untidy and I'm sure I looked lumpy. "Come here, Tubby Coppens," she called, "Don't you know how to dress properly?" She was not calling my sister, who was not tubby and always looked neat. Miss Bodie hauled up my tunic in front of the assembly, pulled down my shirt, and straightened my clothing. I was humiliated by this insensitive and cruel action and suffered the name of Tubby Coppens for the rest of the year; and the sense that I carried too much weight has stayed with me all my life.

On the whole, I was neither happy nor unhappy at school. I know I didn't enjoy it, but it was tolerable. I did well in class and brought home good report cards and stayed out of trouble. I was not hugely popular, nor was I shunned. I was afraid of some of the teachers. Once a week, a terrifying lady we called simply Mademoiselle came to the school, and that day we had to speak French exclusively. She was a large woman with a face like an angry bulldog, and I was in a perpetual sweat in case she fixed those bulgy eyes on me and expected me to converse with her. I think my classmates were as frightened as I was. A lot of our teachers were married ladies. At that time if a woman teacher got married, she lost her job in the public school system.

I lived for the weekends. The rhythm of those days was idyllic for me. My mother was always the same: there for us while we were there. We did everything together, taking our dogs and neighbors' dogs for walks, finding ponds in nearby woods to collect tadpoles in spring. All my happiest memories revolve around those times. And then on Sunday, after tea, we took the streetcar back to school.

During the two and a half years that my father was a prisoner, my mother only received two postcards from him, via the Red Cross. They were undated, so we had no way of knowing when they were sent.

But everything changed in 1945. I was eleven, so old enough to pay attention to the news, and everyone in my class was very tuned in to the war in Europe in particular. D-Day in 1944 had been exciting, with the liberation of Holland, Belgium, and France. The allied forces advanced into Germany. President Roosevelt died on April 12, 1945. Miss Bodie called us all to assembly to tell us about the president's death and we prayed for him. Hitler committed suicide at the end of April and German forces started surrendering on May 8.

Much as we were happy for the end of the European war, my family was more concerned about the Pacific. That part of the war did not end until August 15, a week after the United States dropped the atom bombs on Hiroshima and Nagasaki on August 6 and 9. Pamela and I were at Kewpie Camp, but my mother made a special trip by the Union Steam Ships (Roberts Creek in those days was only accessible from Vancouver by water, although it was only about thirty miles from Vancouver) to show us the telegram she had received. It said, "everything is rosy with Bill," and it was signed by an Andy Yonkers of the US Navy. This telegram had arrived a couple of days after the Japanese surrender. Those must have been difficult days for my mother, not knowing when or what she would hear about her husband, but she never spoke of it and was her usual calm self when she arrived at Roberts Creek.

Pamela and I were excited. I remember I cried but I wasn't sure why. I had a father again. Our immediate question was, when would he be coming to Canada? Of course, no one knew.

As it turned out, the liberation of Indonesia was far from straightforward; some of the islands were not liberated until well into September, and that wasn't the end of it all.

I have not been able to find out exactly how Andy Yonkers or the US Navy got to Batavia and helped my father. Yonkers was an American from Seattle and arrived in Batavia off a US naval ship. There is no record I could find of any US presence in Java at the time of Japan's surrender, and the historical credit is given to Australian forces. But this young American helped my father by sending a telegram to my mother and by providing my father with some very sound advice about what he should do. I know there was a sizeable American naval presence around Indonesia and the Philippines in 1945, certainly around the Philippines. I can only assume that some of their vessels were deployed to provide humanitarian aid to prisoners in Indonesia.

When Japan invaded Indonesia, the native Indonesians welcomed them as liberators. The Japanese encouraged this sentiment by providing educational opportunities for the Indonesians and creating a civil service infrastructure into which Indonesians were hired. In the confusion that followed the capitulation of the Japanese, civil unrest surfaced almost immediately, with Indonesians demanding independence.

The advice Andy Yonkers gave my father was a warning that there was going to be civil war in Indonesia and that he should not stay around waiting for a boat to take him where he wanted to go. Rather, he should take the first opportunity to get out of Indonesia as soon as possible.

My father was in very bad shape in 1945. He weighed ninety-five pounds, had impetigo on his hands and arms, and was afflicted with beriberi. But he acted on the advice he'd been given. The upshot was that he ended up in Australia. My father's adviser had spoken truly. Civil war broke out almost immediately after the Japanese surrender, and on some islands before the prison camps had been liberated. Many of the Dutch people who had survived the war were thrown back into concentration camps for another year or more. Many of these people, suffering deprivation and malnutrition, died during that time. The civil war continued until 1950, when Indonesia gained its independence. The allied forces had been so strapped for administrative control during this civil war that they found it necessary to keep a considerable number of Japanese administrators and soldiers in place to try and keep the peace.

The allied forces in the Pacific had their priorities: namely, to bring their own troops home. The Dutch government had very few resources to provide aid to people like my father. From Australia, he managed somehow to get boat passage to take him back to Europe and to Holland. He arrived in Hilversum, where his parents and two brothers lived, just in time to see his mother die. There had been much suffering in Holland during the war. One of my uncles did a stint in a German concentration camp, accused of helping a Jewish family go underground. My father arrived in the winter of 1945 with nothing. The Dutch Red Cross outfitted him with heavy clothes that he continued to wear for several years after he came to Canada, much to the embarrassment of my sister and me.

In the spring of 1946 he was able to get passage on a boat going to New York. He then traveled by train to Vancouver. He arrived while we were on Easter break. My sister was away. She was spending the holidays with a friend whose parents owned a cattle ranch in Williams Lake. My mother and I went to meet the train. I was very nervous.

My father looked thin, tired, and diminished in his oversized greatcoat and wide-brimmed hat. We were all shy with each other. I of course didn't know him at all, with only the haziest memories of him before the war and pictures my mother had brought with her. He hadn't featured large in our lives even before we came to Canada. He went to the office and the only times we spent with him were on summer vacations when we drove to one of the hill stations. I remembered him as the person who meted out corporal punishment if we had been bad. Misconduct resulted in us getting "the slipper," the object being a rubber-soled slipper of my father's, used only under extreme circumstances on our bare bottoms as we knelt over our beds. Now he was in our midst. I often wondered how he must have felt, coming home from work and being told he had to administer a spanking to either one of us. Slipper episodes were very rare. I remember the threat very clearly to a point where I wonder if the actual slipper was ever administered, because I don't remember ever getting "the slipper."

The evening of his arrival, we sat, the three of us, on the couch in our living room, my father in the middle. He talked about his experiences. It seemed to me he wanted to lay everything out. He talked about the conditions in his POW camp, how they were always hungry, how he got advice about certain especially healthful vegetables from friendly Indonesians who were not imprisoned but who did business with prisoners through the barbed wire. He spoke of how he pried the gold fillings out of his mouth to trade for eggs.

He described the deaths in the camp, usually from disease; how the first person to die had been a friend of his and how terrible it had seemed, but how, by the end of the war, he felt too tired to go to a service when someone died. He talked about how the infection he had on his hands had probably saved his life, because during 1944 the Japanese were taking men from the camps to work building an airstrip on a coral reef off Indonesia. He had not been eligible because of his hands, but he later found out that the ship on which the men were transported had not been marked as containing POWs and had been blown apart by US planes. Also, men taken to build those airstrips often died from infections. Coral is sharp and causes lacerations that can become infected.

He talked about how when you have everything stripped away, you attach value to any small item, an empty tin, a scrap of cloth. And about the pettiness of some of his captors, who would take pleasure in scattering these few treasures that prisoners had collected.

He told us how, after the surrender, he and some other former prisoners were charged with going around to some of the facilities where women and children had been housed. And how much worse he thought they had had it, crowded into insufficient housing, forty to fifty forced to live in a house the size of the one we had lived in. He said how brave he had found those women.

He bore surprisingly little animus towards the Japanese. He felt most of his captors were not cruel men. They were doing their duty. And towards the end of the war when food was very scarce, the soldiers were starving too.

At the end of the evening he made it clear that he now wanted to go forward and did not want to talk about his experiences any more.

And that was the way it was. It wasn't until he was in the early stages of dementia in his late seventies that some of the horrors he had experienced resurfaced.

My father's presence required a great deal of adjustment on all our parts, my mother included. I know what a damaged man he was. At the time Pamela and I just had to bite our tongues and become accustomed to the mantra "Don't upset your father." He was not a strident or violent man. But he did get upset and lose his temper easily. And he was very jumpy and timid, all characteristics of posttraumatic stress disorder.

There was the issue of his getting a job. He was an immigrant, although he was fluent in English. His qualifications from Holland were equivalent to a CPA but not recognized in Canada. All the men who had served in the Canadian Armed Forces were being demobilized, and they were given preferential treatment on the job market. My mother had her job at the Workman's Compensation Board but my father's morale was very low because he was not the breadwinner.

I don't know where the notion came from, but he decided to try his hand at being a door-to-door salesman and signed up with Watkins. Watkins products covered toiletries and over-the-counter drugs of good quality. Like Avon and Fuller Brushes, Watkins products were only available from accredited salespeople. My father purchased a sizeable inventory of Watkins products. They were stored in our basement.

My father was not a salesman. He was a shy man, more introvert than extrovert. He was in a strange land with which he had no familiarity. And he was in fragile health. I don't know how long he punished himself trying to go from door to door selling toiletries. It wasn't long. What I do know is that we had a basement full of Watkins products that took us many years to use up. Our bathroom stocked only Watkins shampoo, soap, and cough syrup until I graduated from high school.

My feelings about my father were very confused. I resented him for having disrupted the cozy life the three of us had. I felt no special kinship towards him. He was not a demonstrative man. I don't recall ever having been hugged by him, or even touched. For that matter, we were not a demonstrative family. My mother kissed us goodnight on the forehead, but she wasn't a hugger either. When I thought about my father and what he had suffered I felt enormous, almost crushing sympathy. In later years, when I read the letters he'd sent my mother before the war, I got a completely different picture of him: warm, humorous, courageous. I realized I'd never known him. The man he had been had was destroyed by the war. Years later, after his death, my mother described the young husband she'd fallen in love with and how hard she'd tried to bring him back.

It was 1948 before my father got a job suitable to his temperament and abilities. It was local, in Kerrisdale, with a small accounting company. Kerrisdale is a middle-class neighborhood adjacent to the Dunbar neighborhood where we lived. My father felt comfortable in the atmosphere and fully capable of doing what was required of him. We moved to a slightly larger house. My sister and I had separate bedrooms, and it had a gas furnace, so we didn't have to fill the hopper with sawdust. All the residual Watkins products moved with us to a new storage place in the basement.

Pamela and I were finally allowed to attend public school. I was in grade ten, Pamela in grade eleven. And we bought a car, a 1935 Buick. My mother brought home a stray dog, a puppy she'd befriended who was hanging around a butcher shop near where she worked. The attendants said the dog had been hanging about there for several days and my mother took pity and brought him home in a taxi. These three apparently disparate events – the car, my father's job, and the dog – are firmly linked in my memory.

My sister and I attended Magee High School, in the same neighborhood where my father was employed. After school we hung out in Kerrisdale with our friends. Many of the customers of the accounting firm my father worked at were shopkeepers in the neighborhood. Not infrequently, my father would drive through Kerrisdale on the way to one customer or another, so it was not unusual to run into him when we'd be coming home from school with our friends.

My father was still wearing the hat and greatcoat he'd received from the Dutch Red Cross. The coat was too big, as was the hat, making him seem even more diminutive than he was. The Buick was a large car. My father, under his big hat, was hardly visible. The car looked like it was being driven by a large hat. To make matters more bizarre, my father had fallen in love with the dog, Busuk (which means bad smell in Indonesian). Busuk often accompanied my father on his trips around Kerrisdale. The dog sat in the front seat, his head resting comfortably on my father's shoulder. If my father saw us, he'd stop the car and offer us a ride home. Being teenagers, we were easily embarrassed in front of our peers, who would wrinkle their noses and say, "Who was that?" When I admitted the relationship, the look of puzzlement wouldn't change. They'd say, "Does your dog always ride in the front seat? Where does your mother sit?" Until I was asked that question I'd never thought it strange that my mother, sister, and I always sat in the back seat when the dog was aboard.

Fortunately the car developed a quite audible squeak that sounded like a high-pitched oy-oy-oy sound when it was being driven. So, like the crocodile in Peter Pan, Pamela and I were alert to the sound and were usually able to avoid contact with our well-intentioned father, just like Captain Hook evaded the crocodile.

I was very happy that we were allowed to go to public school. But I was very nervous. I knew only one person who would be going to Magee and would be in my grade. She had lived across the street from us when we lived on Thirty-Eighth. She had an older sister who was in grade eleven. Their names were Patsy and Peggy Beck. Patsy was older than the other children on the block, and was bossy but well-intentioned. She willingly took my sister in hand. Peggy, the one in my grade, was a bit unusual. The reason that adults on the block spoke about quietly was that Peggy had had a fall when she was very small, had been concussed, and had never been the same again. I suspect she was just a hyper-nervous child, because she was hyperactive and prone, as a younger child, to jumping up and down and having screaming fits. But as a grade ten girl, although a bit unusual still, she certainly passed muster. Peggy hung out with a group of girls from the wider neighborhood. They were girls who had probably attended grade school and junior high together. But Peggy was very kind and helped me get to know people and feel less isolated.

After the first few days at Magee, I settled right in. I must say that although my schooling prior to high school had been chancy, the education I had received did me proud. I learned very little new in English (the same texts we used in grades seven and eight were used at Magee in grades ten through twelve) and French (my grounding from Mademoiselle had been thorough). In those days grammar was taken seriously, and at Queen's Hall sentence parsing had been drilled into us. It was in science and math that I was exposed to new information in high school. And it was in these subjects that I became absorbed, math in particular.

The girls my sister and I were friends with at high school came from homes with fathers who were professional men. These girls had lots of cashmere sweaters. Although there was no uniform, there might just as well have been one – cashmere sweaters with plaid or plain skirts, mid-calf, either straight or pleated. Bobby socks and saddle shoes were what we wore on our feet. The boys wore cords and V-neck sweaters over white T-shirts and had crew cuts.

I knew we were not as well off as my new friends. I had worn a uniform until I went to Magee and had very few clothes. My parents made us responsible for buying our clothes with an allowance of about ten dollars a month. You couldn't buy a lot of cashmere sweaters on that kind of allowance. We made our own skirts, but cashmere sweaters were something I could only dream of. I certainly learned how to watch my pennies.

Although I cared about keeping up with my privileged classmates, I don't think I was very clothes conscious, and I'm sure I was still untidy. Years later I ran into a man who had been in school with me and he said very sweetly that he remembered me as the brainy girl whose head was usually in the clouds and who often had her shoelaces untied or a button incorrectly fastened.

I ran into a girl I had known at Kewpie Camp. I had disliked her intensely then. She was in my hut and was misbehaving one night, swinging on a rafter. She let go and landed on my bed, breaking my badminton racket in half. I must have been about eight then, and I cried because I treasured that racket. She got into trouble from a counselor and had it in for me after that. She and one of her friends would sneer and call me a crybaby and a tattletale. I hadn't tattled. The counselor had found me crying and another girl had told her why.

The racket breaker's name was Dee Edgell (Dee being short for Drucilla), and I thought she was going to continue to be obnoxious at school. Her first comment to me was, "Your zipper at the back of your skirt is undone." But it turned out she was being considerate. She became my best friend in high school. Funnily enough, we both still played badminton and became staunch partners in the game, all animus forgotten and forgiven. We went on to win the BC junior doubles and the school championships every year we were at

Magee. She was a far better athlete than I was and put up with me as a partner out of friendship. I helped her with her math.

Dee was the girl I went to movies with. We talked on the phone every day when we got home from school. We played tennis and badminton at the Vancouver Lawn Tennis and Badminton Club as juniors. We shared our crushes and excursions to something called the Saturday Night Club, a church social that allowed teenagers to dance in the church hall on Saturday nights. The hall was almost totally dark and we slow danced to Glen Miller and other musicians in the steamy darkness, full of hope and expectation. Boys stood on one side of the hall, laughing and emitting loud noises, while girls clustered in groups whispering and giggling on the other. If a boy asked to take you home, that was a coup. Since Dee and I traveled as a unit, sleeping alternately at each other's houses, we had a pact to turn down any offers of an escort home.

Dee and I were loosely associated with the cashmere sweater set that I had met through Peggy. There was never any doubt that we were all heading to university when we graduated. I don't know how many of these girls had career aspirations, but in the middle-class environment of that school the assumption was that a university degree was nice to have but was not necessarily a career training. The reasons were openly discussed when we sat around eating our bag lunches. Universities were where you met boys who were going to have successful careers. Where would you meet eligible men if you went to secretarial training and then into an office? Nursing school was an option, because you could meet doctors and medical students – who were almost exclusively men – but nursing was hard work and not so much fun as university.

I wasn't at ease during these discussions. I was thinking seriously about trying to get into medical school, but I wasn't sure and didn't say anything. I didn't have a defined goal when I was in high school but I knew with absolute certainty that I would go to university and that I would have a career. The thought that I would end up as a dependent never crossed my mind. My need to be self-sufficient was deeply ingrained.

CHAPTER THREE

Where I Wanted to Be

I loved university. I didn't consider going anywhere other than the University of British Columbia. UBC was in Vancouver and tuition was reasonable. My parents agreed to pay for my first year, as they had for my sister. Then it was up to me. We lived at home while we were in school. Pamela went to teacher training school after her first year and became a primary school teacher. I had other plans.

There were about five thousand students at UBC when I attended. After the war, returning veterans overwhelmed the universities. The institutions responded quickly by installing disused army huts all over campuses. They served as classrooms, labs, and residences. Although they were considered temporary, a few are still in use on the UBC campus, sixty years later. In the 1950s, many classes were in huts.

I soaked up the atmosphere, so free from the petty rules that had governed my daily school life. I loved the big anonymous classes. Nobody cared if you attended. I did attend classes, because I was interested. I took math, chemistry, zoology, and bacteriology, as well as English, that first year.

The labs, the hands-on part of science, thrilled me. One of our first labs in bacteriology involved the simple exercise of swabbing various parts of our bodies (face, hands, hair, etc.) and then plating the swab onto agar to see how many different bacteria grew on the Petri dishes we were given. It was a simple demonstration of the ubiquity of microorganisms. The plates were incubated until our next lab.

We got to look at the diverse array of colonies growing, then stain them and examine them microscopically. It seemed miraculous that a round, sometimes pigmented colony could arise from a single microorganism in just a few days. Most of the plates produced typical bacterial colonies: white, yellow, and sometimes pink. There was also the odd mold, often the green of Penicillium. The boy who was working across the bench from me had a plate containing much more growth than the others as well large mucoid glistening colonies that looked like gobs of mucous. This plate didn't look like any others that my fellow students were examining. He seemed embarrassed by what he had produced. Other people at the bench started looking at his plate and commenting on it, asking him what part of himself he had swabbed to produce those horrible looking bugs. He admitted to having swabbed his armpit. Bacteria love warm moist environments full of nutrients. It was in this class that I discovered how much I enjoyed messing around in a lab.

It has never bothered me that we carry more bacteria around with us than we have cells in our bodies. I was hooked by these fascinating little life forms from my first few classes. I didn't mind the reek of growing cultures of bacteria or the strong disinfectant into which we put all the pipettes that we used. I decided that this would be my major.

My sister had a steady boyfriend named Allan Campbell. He was at UBC in forestry. He introduced me to a friend of his named Howard Gerwing, who was majoring in English. We started dating. He later became my first husband.

Howard was majoring in English and history. He wanted to be a writer and had a deep love of books. His best friend was a Byronesque character named Bill Dumaresque who was already something of a poet and had a couple of pieces published in a literary magazine called *Prism*. He always had a coterie of female admirers around him. They wore black turtlenecks and chunky jewelry. I felt outclassed in my saddle shoes and bobby socks, but nevertheless soaked up their arty atmosphere in coffee shops where poetry was read several times a week. There was one we went to regularly, called the Mattress Factory, where we smoked and drank coffee and talked seriously. We were not political, as I recall; I suppose we were too self-absorbed.

I was very happy at university. I felt I had arrived where I wanted to be. I was avid to learn. After my first-year Christmas exams, where I barely squeaked through and got my wakeup call, I realized what studying entailed. I had never studied until then. In high school exams had not been challenging, so I had no study habits when I started university. It wasn't that I didn't like the idea, I just didn't know how. But I learned quickly.

At the end of my second academic year I sought out Dr. Stock, the professor who had taught my first course in bacteriology. I told him how much I had enjoyed the course and asked his advice on courses to take to get into the honors program. To my surprise, after our talk, he offered me a job for the summer in his lab. I felt extraordinarily lucky to be given this opportunity. Perhaps he was giving me a leg up because I wanted to do an honors degree. I didn't think I stood out in any way. There were at least two hundred students in his class and he had offered me the only job he had available. I was ecstatic. I was about to do real research. It was a mundane project, looking at the best medium to use when cryopreserving bacteria. Still, it was research, and the bacterial strain my boss had selected was Neisseria gonorrhoeae, the causal agent of gonorrhoea, because the gonococcus is a very delicate bug and not easy to cryopreserve. The belief was that if one could freeze-dry these fragile organisms in a given medium, those conditions would work for most other bacterial species.

Allan, Pamela's boyfriend, and Howard went away every summer working as tree surveyors for the BC Forest Service in remote parts of the province. They only got back to town a couple of times, proudly displaying beards and smelling of wood smoke and sweat.

I spent my time at the lab and at the Vancouver Lawn Tennis Club playing tennis with Dee, who was working in an office. She was not going back to university in the fall. We had been juniors at the club and were allowed very inexpensive access to senior membership. We still played badminton there in the winter. We met "older men" at the club, young lawyers or other young professionals. My mother was not in favor of "older men." I'm not sure why. These "older men" I think regarded Dee and myself as pure and off limits, whereas boys of our own age were far more prone to be sexually aggressive.

While Allan was away that summer, my sister met a young man from London, Ontario named Gerry McCann. He was large and loud, while Allan was quiet and quirky. Pamela fell in love with Gerry and broke up with Allan. Allan didn't get over that for years. I felt very sorry for him and certainly preferred him to Gerry. Allan and I remained friends until he died in 2014.

UBC in the 1950s was a small provincial institution. In my department, with the exception of Dr. Dolman, no faculty members had ongoing robust research activities. Nevertheless, I was inspired by what I learned through my courses and the reading I did. I knew science was going to be my career, either by enrolling in medical school or by going into honors bacteriology and immunology. I ran into a girl I knew from Magee named Peggy Andreen. She and I were in the class of about twenty students who were majoring in bacteriology. Slightly more than half the class were girls. Graduates from microbiology were sought after as lab technicians, so the probability of employment after graduation was high. She was the only girl beside myself from our year at Magee who had gone into science. We became lab partners that year and both of us received top marks in our courses. She was set on a career in medicine, but I still wavered. I made my decision after Christmas when I realized that I really wanted to be in a lab. I wanted to be looking for answers. I dropped my ideas of medicine and continued in science.

In my final year, Howard and I decided we would get married in the fall after I graduated. He had already graduated and was enrolled in a master's program in Canadian history. I was completing my final year in honors. We would work for the summer to earn some money, then get married and travel to England, where Howard would pursue his history research at the Hudson Bay Company archives. I would seek a lab job as a technician for a year, and then we would return to North America, where I would go to grad school. By then I knew I wanted a PhD and to have a life doing research.

Claude Dolman, then the head of the bacteriology department, hired me for the summer after my third year. I was one of three honors students in our year. The other two students were men. All three of us got high grades in our courses, but the other two had established summer jobs, one in the reserve navy and the other as a surveyor in the interior. I assumed that was why I was offered the position of research assistant. Dr. Dolman had a coterie of female assistants, so I also presumed he preferred female staff.

Everyone was afraid of Dr. Dolman. He was a very austere British academic who seemed haughty and rarely smiled. In our senior year he taught a course in the history of microbiology, which I found riveting. I sat spellbound through all his lectures, which were more like stories to me – stories of how early scientists had made their discoveries, how their observations had led to conclusions. How William Jenner, a bird watcher with an eye for pretty milkmaids, observed that girls working in dairies had such perfect complexions, unscarred by smallpox, a disease rampant in Europe in the eighteenth century. He also noticed that these girls frequently had lesions on their hands that resembled smallpox pustules, and that these lesions came from being around cows that often also had pustular spots on their udders from cowpox. He must have gone away and had a think about these two observations, then put together a theory about the possibility that infection with cowpox afforded some kind of protection against the human disease smallpox. He was jeered at in the press at the time, but within his lifetime vaccination, using preparations from the cow virus, Vaccinia, became widely used to protect people from smallpox. Cartoons appeared in London newspapers of people with cow's heads growing out of their arms. When some members of the nobility accepted the concept, the procedure became widely adopted. It was used as a successful preventive for smallpox until the disease was eradicated in 1967. The term "vaccine" owes its origins to this virus.

I discovered that Dr. Dolman had a droll sense of humor. In discussing the way in which Columbus's sailors brought syphilis back to Europe he commented that while the old world civilized the new, the new world syphilized the old. While there is some question as to whether syphilis did originate as a benign condition in American Aboriginals that either mutated or was far more serious in Columbus's crew, the coincidence of syphilis becoming rampant in the sixteenth century and the return of Columbus is inarguable.

Dr. Dolman's research interest was in Clostridium botulinum, the causal agent of botulism and producer of one of the deadliest toxins known to man. He had a catalog of all of the cases of botulism recorded in British Columbia, a frozen collection of all the foods thought to have been the source of the toxin, and a collection of all the strains of C. botulinum that he had isolated from the foods.

People get botulism from food in which the organism has grown and produced its toxin. Because the bug is an anaerobe (it can't grow in the presence of oxygen), incriminating foods are usually home-canned goods that are neutral in character, like beans. These microbes grow and produce a protein that, upon consumption, can kill at very low concentrations by paralyzing voluntary muscles. First Nations were particularly prone to botulism because they were partial to something called salmon egg cheese, a fermented salty product from salmon eggs – I suppose, a kind of caviar. Solid masses of roe placed in a closed container and allowed to ferment also provided an excellent anaerobic environment for C. botulinum to grow. Claude Dolman had isolated and classified five different strains of botulinum over the years and had made it his business to prepare antisera to all five types so that they could be used to treat afflicted people, so long as the strain was isolated and typed.

The job I was offered was to learn how to grow these bacteria. If these bugs are grown in liquid broth, they produce the toxin, then release it into the surrounding liquid. I had to grow the bacteria, filter the broth containing the toxin, test the toxin in mice, and then treat the preparations with formaldehyde. Formaldehyde altered the toxin molecule sufficiently to inactivate its lethal properties without altering its overall three-dimensional properties. This process is used to convert a toxin into a toxoid. Toxoids can be used safely to immunize people against toxins. My job was to use these toxoids to immunize rabbits against the various strains of botulism and collect blood from them once they had produced antibodies against the toxoids. Cells from the rabbit blood were separated from the clear serum. This material could be tested for the amount of specific antibody and then used if some unfortunate person was diagnosed with botulism. I don't know how frequently Dr. Dolman was called on, because botulism is quite rare.

Cosmetic botox is, of course, botulinus toxin used at extremely low concentrations. It exerts its cosmetic effect by paralyzing the muscles of the face, preventing frowning and other facial expressions.

I worked with two other people that summer: a Miss Chang, who was Dr. Dolman's senior assistant, and a woman called Mollie Enns. I was thrilled with the job and extremely grateful to Dr. Dolman.

He did me another huge favor that would profoundly affect my life. He knew I planned to go to the United Kingdom in the fall and offered to write me a letter of recommendation and introduction to an old associate of his who was a senior fellow at the National Institute for Medical Research at Mill Hill in north London. He was sure I would be able to get work there for a year.

I had been very frightened of Claude Dolman because of the reputation he had both inside the department and outside. When the university decided to form a medical school after the Second World War, Claude Dolman expected to be made its first dean. He had obtained his medical degree from St. Mary's Hospital in London and trained in research under Sir Alexander Fleming, the discoverer of penicillin. He had come to Vancouver in 1936 to take the chairs of both bacteriology and the Department of Preventive Medicine. In 1941 he also assumed the chair of the School of Nursing. It is not surprising that he expected to be made Dean of Medicine. He was well qualified. But internecine university politics intervened and he was denied what he assumed was his rightful post. What transpired is shrouded in gossip and ill will. I know he was cordially disliked by many of his peers. None of his faculty ever spoke to him. I have said already that he was autocratic, his attitude superior and harsh. And he was a snob. His bitterness about being passed over resulted in his refusal to allow the bacteriology department to become part of the nascent medical school and his isolating himself from his former colleagues.

The honors program in the department required that honors students take extra courses over and above the requirements for students graduating with a bachelor's degree. For me to graduate in the spring following my fourth year I had to take seven courses. One of these was a seminar and another was my research project and the thesis I had to write. I wanted to graduate in the spring because I could apply for graduate scholarships that were awarded after spring grades came in. Also, I had to work during the summer. I needed to save money, as Howard and I planned to marry in the fall and travel to Europe, so I didn't want to spend the summer at summer school. To take an extra course I had to get permission from the Dean of Arts and Science. The university had no faculty of science in those days, just a department of arts and science, even though all the courses I took in my third and fourth year were science courses.

The dean at that time was Walter Gage, a professor of mathematics. Dean Gage was reputed to be a man of great good will, who never forgot a student's name and went to bat for them. The rule about numbers of courses taken was frequently broken, so it wasn't unusual for students to take twenty-one units, particularly if one course was a thesis. Dr. Dolman wrote a request to Dean Gage that I be permitted to take the extra units. Dean Gage refused, even though I was a first-class student.

I made an appointment and went to him to argue my case. He refused again, citing the rules. Dr. Dolman called him and reiterated his request, explaining that he, Claude Dolman, would take responsibility. Dean Gage refused again. So I had to register for my thesis during the summer so as to graduate in the fall. Since I had already completed my thesis, I was free to work that summer. Had I been permitted to take the course, I would have graduated with one of the top grades in the year in science and won a number of prestigious scholarships. Instead, my degree was awarded in November of that year, when I had already left Canada to pursue a higher degree in the United Kingdom.

I was furious at the time. I knew other students who had been granted permission to take extra courses, and they were weaker students than I. Dean Gage was a bachelor. Boys joked about getting passing grades by going to the dean and charming him.

I grew up fully accepting that men had privileges that women did not. It was a fact of life for women in the fifties. Men were the breadwinners, women the homemakers. Walter Gage granted favors to men that he did not to women. He probably thought that universities should be the sole domains of men, for all I knew. But the blatant unfairness and disregard for ability seemed utterly shocking to me at an institute of higher learning. There was no one to complain to. He was the dean. It was my first experience of blatant sexism that overrode reason. And I can still feel angry about it.

But I was probably lucky that I had chosen the biological sciences as my major. Women were welcome in this field. Women made up the majority of hospital lab technicians and research assistants in institutions, and many of them had degrees in microbiology or biochemistry.

If I had chosen to go in for engineering, architecture, forestry, or other applied sciences I would have come across more gender bias.

My parents, during the years I was at UBC, showed very little interest in what or how I was doing and only passing interest in what I was intending to do after I graduated. I was hurt by their apparent indifference, but didn't speak of it until years later when I asked my mother why she had shown so little interest. She was quiet for a time. She clearly understood that she had disappointed me. Then she looked at me and said something along these lines:

"I don't know," she said. "I should have shown more interest because I was curious and impressed by what you were accomplishing at UBC. What you were doing was over my head. And I didn't want to show favoritism over your sister. I suppose I was still trying to take care of your father and hold everything together."

"Well, at least you didn't stand in my way or try to steer me. I got to do exactly what I wanted," I said.

CHAPTER FOUR

Life in Bed-Sits

Howard and I got married in September of 1955 and boarded a train that took us across Canada to Montreal, where we set sail for Liverpool on the Empress of France, a liner owned by the Canadian Pacific Railway. Originally christened the Duchess of Bedford, the ship had put in heavy service during the Second World War as a troop ship. After the war it served as transit for returning soldiers and war brides. It underwent a refit in 1949 and reappeared as the Empress of France.

I had a mental picture of romantic transatlantic travel on ocean liners from movies in the fifties, and the Empress of France was close enough to those portrayals to meet my expectations. We traveled second class and had a tiny double cabin quite far down in the ship. But the dining room was elegant, with a dance floor in the center of it. Every night a small group of bored-looking musicians played contemporary music for us to dance to. The Empress of France, second class, was a down-market version of glamorous scenes with Fred Astaire and Ginger Rogers whirling around the floor.

We landed in Liverpool and took the train to Euston Station, where my aunt Marjorie, my mother's younger sister, met us. She and my mother were very close and maintained a weekly correspondence all their long lives – they both lived into their nineties. I felt I already knew her because my mother would read excerpts of her letters to us. My mother developed age-related macular degeneration when she was in her eighties and her vision declined rapidly. At that time, they learned to use tape recorders so they could continue their correspondence with cassettes.

I recognized her immediately because of her resemblance to my mother. She took us back to her tiny apartment in Kensington. In the main room was a single day bed on which her son Stephen slept when he came home on weekends. The apartment was slightly bigger than a bed-sitter, because it did have one closet-like bedroom where she slept and an alcove with a stove and sink. The bed sitting room or bed-sitter was standard accommodation in London in the fifties and sixties. Bed-sitters all contained the ubiquitous day bed, shrouded in a heavy spread, that served as a couch and a bed, with odd assortments of furniture to accommodate eating and sitting around. Kitchen facilities were minimal, an alcove with a gas stove and perhaps a sink, along with the shilling-in-the-slot gas heater.

Stephen was doing his compulsory military service in the Air Force at that time. That first night, Howard and I shared the day bed. It was fiendishly uncomfortable, so we started looking for a bed-sitter ourselves the following morning. We knew we only had a couple of days before Stephen came home for the weekend.

Marjorie took us to the local post office on Kensington Church Street, where there was a billboard advertising locally available rooms. We found a basement bed-sitter on Argyle Road, close to Kensington High Street and Kensington Gardens. Our landlady was a pleasant older woman called Miss Squire, who lived with her slightly disabled sister on the main floor and rented the upper three floors out as well as the ground level. Miss Squire usually rented only to single ladies. She told me this later, saying that she relented when she saw us because we were so young and innocent-looking. As Canadians, she found us quaint. And she also realized that it might not be a bad thing to have a young man around the place.

The house was a typical Victorian row house with a tiny garden plot at the front and steep steps down to ground level where the garbage cans sat. A door by the garbage cans opened into a long hallway that extended to the open staircase to the main floor. The hall continued on from the stairway to the back door.

Our bed-sitter occupied the front half of the ground floor. It was a large room, furnished with two day beds, a couple of chairs and a table, a chest of drawers, and a large wardrobe for storing our clothes since there were no cupboards. There was a little alcove at the front that had a washbasin in it. There was orange mold growing on the wall under the sink. The toilet itself was just outside the back door, in a kind of attached shed, and very unpleasant on cold nights. I must admit I resorted to using the washbasin on occasion, unwilling to get completely chilled by going outside. The windows at the front looked out on the garbage cans and up to the feet of people passing on the road. We were permitted two baths a week, which we took in the bathroom on the main floor. We took baths on different days and shared the bathwater so that we managed to get four baths a week.

The only other room on this floor was the large old kitchen that took up the back half of the bottom floor. A wide double door led from the hallway to the kitchen. On the other side of the hallway, under the slope of the stairway, was the scullery with a large sink. In the kitchen itself was a regular four-burner gas stove with an oven. Along the wall between our room and the kitchen was a counter, above which were shelves. In the center of the room was a big old table with four chairs around it. The kitchen also had a coal-fired water heater and the water tanks. Other tenants hung their washing on a rack above the water heater. They did their washing in the scullery. The water heater gave off quite a lot of heat and this became a very important feature as we got into winter. As Miss Squire was explaining all this to us she said with great pride that they also had a mangle that the residents used. I had no idea what a "mangle" was, but it sounded brutal. I discovered later that a mangle was a hand-turned wringer.

Miss Squire told us we could use this common area to cook, on the understanding that other tenants were free to access the space when they wanted to do their laundry in the big sink in the scullery. Washing machines were virtually non-existent in post-war England at that time, and laundromats rare. Washing was done by hand. We had no problem with other tenants sharing the space. There was a kind of pantry that extended out behind the house where we could keep our food. It was a cold clammy place, but since there were no refrigerators it provided a better way of storing food than in the kitchen.

We were only a few blocks from Marjorie and Stephen's apartment and quickly became very close, having meals together and going out to the pubs, galleries, movies, or the theater. When I cooked for us we ate off the kitchen table in the common space, cozy and warm beside the water heater.

Marjorie had an office job in Bloomsbury. Although her husband Paul was around and they were friendly, they did not live together. Her marriage to Paul, years before, had not pleased her parents, who thought Paul was an unreliable rogue and wastrel – which he was, but he also could be charming. Marjorie had two children: a daughter, Diana, who had emigrated to Canada several years earlier and now lived with my parents, and Stephen.

Paul, who had joined the army when the war broke out, left Marjorie with two small children and no support in London, having neglected to arrange for her to get the family allowance from the military. People with children were being encouraged to leave London when the blitz started, so she went back to the family home in Torquay, in southwest England. She became the de facto housekeeper for her parents for the duration of the war. As soon as the war ended, she found her parents another housekeeper and escaped back to London.

Stephen was a talented artist and after his service in the Air Force he formed a successful graphic arts company with three other artists in Soho. It was fun to go to art galleries with him and see works at the Tate and other galleries through the eyes of an artist.

I arranged an interview at the National Institute for Medical Research, which was in Mill Hill, a village that had become a suburb of London. I nervously made my way by tube to north London to the end of the Northern Line and then by bus to a remarkably pastoral scene of fields and cows and the enormous brick building that was the National Institute for Medical Research. The Institute was on the edge of the Green Belt that surrounds London.

I was interviewed by three people that day. We sat around a large table in a formal boardroom and talked about my education, my intentions, and my interests. One of the interviewers was a woman called Rosalind (Rozzie) Pitt-Rivers. I don't remember who the two men were, but Rozzie impressed me. Middle-aged and quite petite, she had a humorous, slightly simian face. Her gray hair was short and straight and held back by a black ribbon that she wore like a little girl, with a small bow on top of her head. She was a chain-smoker and had the voice to go with it. But she put me at my ease. She was very well known in endocrinology circles for having isolated tri-iodothyronine, an important constituent of the thyroid hormones.

When I told my interviewers that I intended to return to Canada and pursue my PhD, Rozzie said, "Why don't you do your PhD here? We have an available fellowship."

I was stunned and delighted. I had hoped, at best, for some kind of routine technician job if I were lucky, and now I was being offered the opportunity to do my doctorate. I could only conclude that Claude Dolman must have written a glowing letter about me to his friend. And I knew instantly that this was what I wanted to do. I was sure I could convince Howard that this was a good idea. When I broached the idea to him later that day, he was enthusiastic.

My fellowship didn't start until January. It was late October by this time, so I had a couple of months to get to know England and get acclimated before I started on my next big adventure.

We had saved a fair amount of money in Canada, a couple of thousand dollars, which was a lot in those days. I made sure we put sufficient funds for our return journey in a bank deposit that we couldn't touch. Our rent was 4 pounds a week ($20.00). My student stipend paid 500 pounds a year. Howard would now have longer to complete his research at the Hudson's Bay Company Archives, so he'd be able to get a job. He got one at the Kensington Public Library that had flexible hours so he had time to go to the archives for his research.

I loved the novelty of London. I read the *Forsyte Saga* and *1984* during those weeks before I started at the Institute. It was thrilling to realize I was actually close to where both these books were set. I could take a bus to the place where the bookstore described in *1984* was located, near Paddington Station. I loved walking those gloomy November streets accompanied by these fictitious characters as well as famous people from the past identified by the blue plaques on houses showing where they had once lived.

But I was unpleasantly surprised by how cold I felt once winter arrived. The big old houses retained the damp and the little gas fire we had in our living space threw heat for no more than a six-foot radius. So we spent our time either huddled around the gas fire or sitting in the common kitchen area by the coal-fed water heater. The gas fire consumed shillings by the bucket. Since most Londoners survived using these shilling-in-the-slot gas fires, by the time that meter readers came around at the end of the month shillings were in very short supply. Miss Squire allowed us to open the moneybox and reuse the same shilling over and over again. All the gas fires in the house were separately metered, so we simply paid the gas bill for our meter. That was very kind of her, because many landlords pocketed the difference between the intake of shillings and the actual gas bill. I became accustomed to keeping my overcoat on after I got home until the gas fire took some of the iciness out of the air. London still had bad fogs, known as "pea soupers." The fogs permeated houses and left slimy dirt on the furniture.

At Christmas, we went down to Torquay with Marjorie and Stephen to stay with my grandfather and to meet other members of my family. Torquay is on the ocean in the south of Devonshire and is a resort town with many grand hotels, smaller hotels, and untold numbers of bed and breakfasts.

My mother had grown up in the house my grandfather still occupied. It was a big early Victorian with three full floors and must at one time been quite gracious. Located on a triangle of land not far from the center of town, it had once had a pretty garden. But as my grandfather's car dealership grew, parts of the house became cannibalized to accommodate his growing business and outbuildings replaced the garden. The main floor had been taken over except for the dining room and kitchen. His living quarters were on the second floor, where the comfortable sitting room and Grandpa's bedroom occupied the two front rooms. There were two big bedrooms behind, and an enormous bathroom with a fireplace in it. There were more bedrooms on the third floor.

Grandpa was ninety when I met him, and although his vision and hearing were failing he was still fully alert and ran the business. Shortly before our arrival in England, my mother's brother, Arnold, who was managing the business with Grandpa, ran off with an employee's wife, a Mrs. Revel, causing a scandal that reverberated in the Torquay business community. When Arnold fled he took a large sum of money with him. To avoid crippling death duties, Grandpa had signed the business over to his son a few years earlier, so Arnold had absconded with money that was nominally his. But what he had done hurt the business significantly and ultimately created difficulties for my mother's and aunt's inheritances.

So now, Grandpa and his grandson Tony (Arnold's son) were running the business. Tony, my cousin, had served in the military in India during the war and had returned hoping to go to university, but he had been pulled in to work in the business, which he never liked.

Tony was about ten years older than I. He was the nephew my mother had taken care of when he was little and he still retained the warmest memories of her.

I had grown up with only my mother and sister as family, until my father returned from the war. Even then, we had no extended family in Canada. Now in England I was discovering relatives for the first time. They had embraced me and made me feel that I belonged. It was a novel and heartwarming feeling.

CHAPTER FIVE

Woefully Outclassed

I started at Mill Hill in January, 1956. My research project was multi-disciplinary and I was assigned three supervisors. Rosalind Pitt-Rivers, or Rozzie, whom I have already spoken of, was one. A biochemist called John Cornforth was also assigned. He was profoundly deaf and worked closely with his wife, who was also a biochemist and who served as his interpreter. He could easily read her lips and she understood his muttered words. John received a Nobel Prize in 1975 for his work on enzyme structure and function. He was later knighted. He was a warm and generous person. I got used to the communication via his wife. He was always willing to stop and chat, via his wife, and answer any questions I might have.

My lab space and desk were in the lab of David Long, my other supervisor, a man with a research degree in medicine and a background in immunology and microbiology. He was a big heavy-set man with thin oily hair and an exceedingly large nose.

In England, graduate school did not involve any course work. This was in sharp contrast to graduate programs in both the United States and Canada, where most graduate students spent at least their first year in graduate school taking courses. In the United Kingdom the candidate was assigned a scientific question and given the task of answering that question through experiments. Supervisors were assigned to give advice and guidance, but not to direct the research project itself. It was very much a sink or swim environment. It was up to the candidate to review the literature, design the experiments, and accumulate relevant data.

I was offered a specific project. It was an uninspiring one, but would school me well in research training and scientific thinking. The British attitude about graduate school was that projects served as a training ground, research should be well outlined, and appropriate results should be achievable. Projects with a lot of uncertainty about outcomes were not approved of. You were completely on your own to design and carry out your experiments. Risky research was what you did after your training.

My project was to study the effects of stress on immune responses in different animal species. There was some published evidence suggesting that responses to stress differed between animals that don't need vitamin C and those, such as we humans, that require it to be part of our diet. My project was to study the effect of stress on a number of immune response parameters and to determine if the requirement for vitamin C affected those responses. It was a very suitable project for research training. I learned many new techniques and how to analyze data and run statistical analyses. I had to construct carefully controlled experiments and integrate interdisciplinary research.

There was another woman in the lab named Jennifer Shewell. She had completed her PhD at the Institute in David Long's laboratory and was staying on to do post-doctoral research. Her father was a Harley Street physician and she was very patrician. She had recently married a young lawyer and was in the process of trying to get pregnant. I liked her, and she was very kind to me and introduced me to her other women friends within the Institute. These included a woman named Delphine Parrot, with whom I became very friendly. She was probably in her early thirties and was a career scientist at Mill Hill in the department run by Alan Parkes and his wife, Dr. Deansley, giants in the field of reproductive endocrinology and Nobel Laureates. Another stand-out in that department was Audrey Smith, whose discovery that sperm could be cryopreserved using special media was instrumental in revolutionizing the field of artificial insemination. This involved collecting sperm from animals to be used in artificial insemination. Delphine described in detail the procedure used to get a bull to ejaculate in such a way that one could usefully save the sperm released. The hapless bull was tied in a stall while an ovulating cow was paraded in front of him. I developed an immense respect for

Audrey Smith. Both Dr. Deansley and Audrey Smith were formidable women in their fifties. They represented that early generation of pioneering women who persevered in getting scientific training in the 1920s and 30s and went on to have successful careers in research, our true role models. They were fierce and impressive.

Delphine taught me a number of sophisticated surgical procedures and regaled me with stories of what it was like being a graduate student at the Institute during the war. She had been very poor, having come from a working-class background, and attended university with meager scholarships. She once made a bet with someone that she would eat a mouse. And she did. She brought a small frying pan to work and a pat of her butter ration. She killed a mouse, skinned and gutted it, then fried it and ate it. The bet paid ten pounds, money that kept her in real food for weeks. Delphine also told me that people would try to justify running experiments on rabbits during the war that required the sacrifice of normal (untreated) controls, so that they could take the meat home. She said that even guinea pigs were fair game and that they tasted "rather like chicken."

Another friend I made there was a PhD student called Joan, but I don't recall her last name. She was a research assistant in the lab of Dr. Archer Martin. She had just started her work when Archer was awarded a Nobel Prize for his work on partition chromatography. The story had it that Archer had his eureka moment when he was in the pub just down the hill from the Institute. He was with some colleagues and they were engrossed in a scientific discussion. Archer was writing on a paper napkin with his pen. Some beer spilled on the table and seeped into the paper on which he had been writing. As he watched the ink he noticed that instead of a single color blurring with the moisture as it spread, individual components of the ink were separating and forming discreet bands of color. He concluded correctly that individual components of the ink had different binding characteristics with the cellulose, the main constituent of paper, and that this differential binding caused the components to separate during diffusion. The whole science of column chromatography was born in that moment. All the early materials used for the purification and separation of proteins were made up of specially treated cellulose.

There was a big canteen on the top floor of the Institute where everyone went for lunch. Although there were no defined seating restrictions between technical staff and scientific staff, there was an unspoken rule. Most of the technicians were young men and women who had gone straight from high school into a technical training program without attending university. One of the technicians in David's lab was working towards his BSc at night school. But there was a class distinction between them and us.

We scientific staff sat on one side of the dining room and the technicians sat on the other. There was also a coffee room to which the scientific staff repaired after lunch and to which technical staff was not admitted. The man in whose lab I was working, David Long, was a conservative snob. He made a comment the first day I was there that at least in the coffee room we did away with this pretense of egalitarianism. I started disliking him at that moment.

I soon became conscious of British class distinctions. At that time, your accent gave away your position in society and what kind of school you had gone to. David Long and Jennie were obviously "public school" and spoke elegant upper-class BBC English, whereas people like Winnie, the technician I had been assigned, and Delphine were definitely "grammar school" and had less posh accents. As I honed my ear to these accent differences, I realized that among the senior scientists at the Institute there were a number who had not come out of the "top drawer" but had made their way by sheer ability. Today, one rarely hears the overly cultivated traditional upper-class way of speech, and a majority of people in the south of England speak with the standard BBC English, which has a strong London tone. But I was glad at that time that my accent could not put me into a slot. I found myself drawn towards those people who came to Mill Hill by virtue of their talent rather than privilege. They seemed more engaged in the politics of the day and inclined to the left. These people helped me become politically aware.

It was at lunch and tea that I got to know many of the truly great people at the Institute. It was through their examples that I developed my own attitude to research and the respect that I hold for the discipline of science. I was surrounded by people who were at the peak of British science, and making seminal contributions to scientific knowledge. But as I remember them, these eminent men and women did not put on airs of any kind. I would say that humility was a common feature of many of these truly accomplished people. They did not see themselves as greats.

Archer Martin had recently won his Nobel when I arrived at Mill Hill. He would sit with us at lunch in his crumpled pants and shirt (he had been widowed and never bothered to have his clothes ironed) and make a meal of corn flakes, something he apparently ate three times a day. He was a middle-aged man with intense brown eyes and a way of getting distracted during conversations. There was no aura of superiority about him.

I kept my head down and took in everything around me. I felt woefully outclassed, which I was, and a bit of a fake some of the time, which I was not. But the atmosphere at Mill Hill was wonderful. The only pressure put on anyone was the pressure from within, and I felt plenty of that.

The only bleak spot in my new life was David Long. I had only been at the institute for a couple of months when he made a pass at me, grabbed me from behind, putting his arms around me and pressing me against him and kissing me on the neck, saying how I was driving him crazy, I was so beautiful. I was stunned. But I didn't push him away. I just stood there, paralyzed by surprise and thoroughly confused. I hadn't been expecting anything like that and wondered if I had somehow encouraged him. I didn't know how to react. My desk was in his office. I had been going over some results with him. His advances had come from nowhere. Had I led him on without knowing it? I knew I wasn't a flirt. I was twenty-one and he was a man in his forties. I didn't want to embarrass him or make him look foolish. He was my boss and I was on probation. Fortunately, someone came into the lab outside his office door and I was able to get away from him.

I had time to think things through during the underground ride home that night. I would have liked to talk things through with Howard. But what was the point? He couldn't help. We'd probably just have a fight. He'd want me to quit and I wasn't going to do that. I was in a jam. David was my immediate supervisor. Would he fire me as inadequate if I rebuffed him? I was on probation for six months. I dearly wanted to stay at Mill Hill. The thought of David firing me terrified me. I had only been there a couple of months and didn't feel I could go to either Rozzie Pitt-Rivers or John Cornforth and ask if I could move into their labs. And there was certainly no one else I could go to. He could easily say I didn't measure up. I had already experienced unfair treatment from Dean Gage. My only avenue would be to try to avoid any situation that would provide an opportunity for David to approach me. Or perhaps he would lose interest. I hoped he would.

But he didn't give up. He seemed to think he had a right to pursue me any time he got me alone. I didn't know how to deal with him and was afraid to make a fuss. I tried to stay polite, brushing his attentions aside by joking, trying never to be offensive. But I also made every effort never to be alone with him. He particularly liked to trail me into the animal facilities, where I would go to check on my experimental animals. He would follow me behind some of the racks of cages and try to grab me.

I decided to make sure my technician Winnie accompanied me when I went to the animal quarters, on the pretext that I needed help handling the animals. I escaped to the library whenever I could, towards the end of the day after the technicians left. I'd see him coming and my heart would sink. I spent a lot of time figuring out how to make myself untouchable.

After I had put up with David for a few months, he went on a short trip to America and I was left in peace. I found the occasion to tell Jennie Shewell about David. She laughed and said she thought that was happening. She told me how David had gone after her before she was married and how her marriage had not deterred him from going after her as soon as she came back from her honeymoon. She commented on his penis as being "as big as an old boot," which made me realize he had been more successful with Jennie than he had been with me. I told her what I had been dealing with and she suggested I move my desk in with her so that I wouldn't be alone when I was working at my desk. Fortunately, David decided to stay longer in America and Jennie and I were left in peace.

I learned from Jennie that this kind of conduct was very common in a number of the labs at Mill Hill. There were several men known to be lechers, David being one. There were any number of affairs going on between supervisors and students or research associates. These situations rarely ended well for the young women involved. The all-powerful men rarely left their wives, and the women were frequently shamed by other women. Delphine told me about several situations where women being harassed sought help from other women, who supported them in any ways they could.

Before David left, I inadvertently caused him an embarrassing moment, which I don't regret. One day at lunch the conversation had come around to melanomas, the most serious form of skin cancer. David had been on a melanoma ward when he was a medic at Guy's Hospital. I have always been deathly afraid of melanoma. I'm fair-skinned and prone to develop moles, and a good friend of mine (who had my complexion) had died from a melanoma on his back. David made a point of saying that, frequently, melanomas resulted from moles that developed under finger or toenails. That night when I was taking a bath, I noticed, to my horror, a small black mark under my middle toenail. I was so shocked that I sat in the bath for a long time, wondering how long I had to live. I stewed over this for several days, too nervous to talk to anyone about it. Finally, I casually raised the subject with David, asking his advice about whether I should make an appointment to have it (and presumably the nail) removed, if not my toe, foot, or leg.

David immediately assured me that he would make the arrangements for me. He was friendly with a specialist at Guy's who would look at it. David duly pulled out all the stops and made an appointment. He even accompanied me to Guy's. I was seen there by a very elegant man, who wore morning dress under his lab coat, and who peered at my toe and then asked kindly how I'd even seen anything so small. I muttered something about magnification under water. He told me not to worry, that he should see me again in a couple of months to see if there had been any change.

I felt better, after having been prepared to have at least my toe amputated. I remember that Christmas of 1956 intervened, because I was in Torquay. Again, when I was taking a bath there I examined my melanoma and was puzzled because it seemed to have moved. It turned out that my "mole" was actually a mark in the nail, and by the time I was supposed to see the doctor again I had successfully removed the mark by cutting my toenails. When I told David this, I could see that he was terribly cross. He said huffily that he'd ring the doctor and cancel the appointment. A short time later, I told Jennie about my experience and mentioned the doctor's name, which I no longer recall. She looked at me in amused shock. She told me that David had made an appointment for me to see one of the Queen's physicians. We both had a good laugh about having caused David such embarrassment.

But I was not too downcast about my dilemma with David. Howard and I discovered England, and London in particular, on weekends. We would find a name on a tube station that we found interesting and set off there, with no inkling of what we'd find. Names I recall being fascinated by were Crystal Palace, Tooting Bec, and Wormwood Scrubs. We didn't think of trying to find out what they were before we went. There was no internet then.

Crystal Palace was of course the site of the palace built of cast iron and glass for the Great Exposition of 1851. Originally housed in Hyde Park, the building was moved to south London and re-erected in 1854. It was destroyed by fire in 1886. When Howard and I got to Crystal Palace, all we saw was a tangled and overgrown ruin amidst suburban sprawl, the cast iron of the building's skeleton still visible. Tooting Bec was just an ordinary suburb in South London and, of course, Wormwood Scrubs was a huge and ominous prison.

I became pregnant in 1957. That spring my grandfather died at ninety-two. He died peacefully in his sleep. We went to Torquay for the funeral. Marjorie suggested I might like to take some mementos for my mother from Brunswick House. I selected a very old *Encyclopedia Britannica*, a first edition of *Lawrence of Arabia*, and a gold-plated Victorian traveling clock that Grandpa had made and given to his wife. It was a beautiful delicate thing that chimed annoyingly every fifteen minutes, so long as it was wound up every few days.

It was during this time that I became politically aware. Resistance to the atomic bomb was growing as more and more was learned about radiation damage. Britain was a good place to learn about the labor movement. I became aware of the disparities between the working classes and the middle class. I saw what the Labour government of Atlee had done by bringing in National Health and how that program had made it possible for some people to have dental care for the first time in their lives. And I met young people at the Institute who were strongly Labour and I felt myself very aligned with them.

My son Nicholas was born on November 24, 1957. I had pretty much completed the research for my PhD by then. There was a little more to do, but the end was in sight. I started writing my thesis while I was on maternity leave. We had had to move away from Argyle Road and find a more suitable place for a baby. I also wanted to be closer to the Institute. We found a flat in the village of Mill Hill itself, cheaper than Kensington and larger. It was the second floor of a two-storey house that had been converted to a flat. The landlady, an elderly woman in a wheelchair, lived downstairs.

Nicky was a treasure of a baby. He slept through the night after a couple of days and was incredibly easy. We carted him around on the tube when we took our weekend trips back to Kensington to visit Stephen and Marjorie.

Nicky was born at University College Hospital, which had a first-class maternity unit. I believe the United Kingdom was ahead of North America in prenatal preparation and care. I had a rigorous set of very integrated appointments at the hospital throughout my pregnancy that included exercising, lectures on breastfeeding, parenting skills, and so on. Deliveries, even in hospitals, were performed by midwives, unless there were complications. Very few second babies were born in hospitals at that time, the midwives going around on their bikes to do the deliveries. These women were highly trained in obstetrics and perinatal care and engendered absolute confidence in me. I was attended by a midwife and a medical student for Nicky's birth.

Caring for a baby was hard work. There were no disposable diapers and there was no washing machine in the house where we lived. Diapers were boiled and washed by hand and then hung around the little gas fire in the living room to dry. The alternative was to pack all the washing up, pile it onto Nicky's pram, and walk to the village laundromat. I don't recall the laundromat having dryers, so we'd wheel the wet washing back and hang it around the gas fire to dry. If it wasn't raining I could use the clothesline outside.

I went back to the lab at the end of February after I found a nice grandmotherly lady who lived in the village to care for Nicky during the day. I would take him over to her house in his pram in the morning and then take a bus to the Institute. I finished the research I needed to do, handed in my thesis, and prepared for my defense. I had verified the species difference in stress responses but I was not interested in pursuing the topic. The research had provided me with an excellent vehicle to learn how to construct experiments and to test theories. Writing my thesis had taught me how to dig into scientific information and to evaluate that information. I would publish three papers from my thesis, all of them in good peer-reviewed journals. But immunology was still my main interest.

The immunology department at the Institute was full of brilliant people. John Humphries, who had worked with Alexander Fleming on his discovery of penicillin, offered me a post-doctoral position working collaboratively with Rodney Porter. I was delighted with this opportunity. Rodney Porter's groundbreaking research on the structure of the antibody molecule was taking place at that time. He would go on to win a Nobel Prize in 1972 for this work. Antibodies were still a mystery in the 1950s. It was known that they were large proteins, that they constituted a significant percentage of the proteins found in blood, that they were made by the body in response to infection, and that the antibody produced bound specifically to the antigen that had stimulated its production. Most of the common substances that stimulated immune responses were also proteins (antigens), so binding of antibodies and antigens involved protein–protein interactions and a complex array of molecular charge effects. I became very interested in understanding the nature of that binding.

I passed my thesis defense with no difficulty and would happily have stayed at Mill Hill for another couple of years in the immunology department, but Howard was getting impatient to return to Canada. He had only planned on being away for a year before returning to grad school, and had stretched that to nearly three. Our marriage had been under some strain for a few months and I suspected that Howard was feeling frustrated by his own uncertainty about his future.

I got a letter from Claude Dolman offering me a junior faculty position back at UBC, so, regretfully, I accepted and we planned our return journey. I was twenty-four.

My years in England at Mill Hill had shaped my thinking and attitude to learning. Apart from David Long's behavior, at no time did I feel there was any gender bias there, because even though most department heads were men there were plenty of academic women around to serve as role models for me. When I think of all the great men and women I met there, I'm grateful for the wonderful example they set for me as I learned how to think about science. Their openness to other people's ideas and willingness to listen and their laissez-faire attitude to graduate students allowed me to make mistakes and learn by those mistakes. I believe that my whole approach to teaching and working had its roots in that experience.

CHAPTER SIX

Whose Lab Do You Work in?

There were a lot of things I had missed about Canada, and in many ways it was good to be back. We stayed with my parents until we found an apartment in Kitsilano, a Vancouver neighborhood half-way between the university and downtown. We bought some rudimentary furniture, mainly from seedy second-hand stores on Main Street, where I developed my taste for antiques. It was fun to hunt around in dusty crowded corners, hoping to find some hidden gems like art deco lamps or Persian rugs.

My salary was $4,800 a year, which seemed like a lot to me. Howard would concentrate on getting his research finished and written up. He had sacrificed time for me and I wanted to repay him for that. He took a teaching assistantship in the history department.

Many of our old friends were still in Vancouver and we soon had a lively social life. Several of us decided to rent a communal house together. We found a big old house with three floors and twenty-one rooms. It had been subdivided into apartments and had kitchens and bathrooms on each floor, but the suites weren't self-contained. Howard and I took the second floor. Residents on the third floor had to pass through our central hall to get to the stairs to their apartment, but they were friends and saying hi to them as they passed our kitchen doorway was fine.

This house remained a communal living space long after our group left and became known as the Peace House, an icon of communal living in the sixties.

The house was very run-down so all of us worked together, painting all the rooms white and doing minor carpentry work to make it livable. Howard's father was a builder and contractor and helped us a lot.

It seemed strange to re-enter my old department as a faculty member. Claude Dolman, the head of the department, was very gracious and set me up with a lab and an office. We talked about what kind of research I wanted to do and I told him I was particularly interested in basic immunology, studying antigen-antibody interactions. He suggested I might consider bacterial toxins as model antigens for that research.

I knew he was hoping I would want to work on the botulinus toxin. I didn't find the idea unappealing. After all, I had worked on the toxin already, knew how to grow the bugs and produce crude toxin preparations. The toxin itself was strongly immunogenic and could stimulate the production of antibodies. And the toxin was a very interesting molecule. It had never been purified or characterized in any way other than by its biological activity and its immunogenicity. Claude Dolman had no particular interest in such work. He was a classical bacteriologist with a clinical bent. His interests were not at the molecular level, as mine were, but rather in classifying subgroups of microorganisms based on their taxonomy. But he did like the idea of someone like myself broadening the field to which he had dedicated himself.

Antibodies are protein molecules and I knew I was woefully uninformed in protein chemistry, although I had read extensively. So, the first couple of years I was back in Vancouver, I audited a number of graduate courses in biochemistry and chemistry. My British education had been superb in some ways but I needed to bolster my knowledge base. The courses I took included protein and nucleic acid chemistry and physical chemistry. Because of my interest in antibodies, a deep understanding of proteins and protein–protein interactions could only help me to do better scientific research.

A new biochemistry faculty member, Gordon Dixon, taught the protein chemistry course. He was an asset to the biochemistry department. He had done his PhD under Oliver Smithies, another British scientist who, like Archer Martin, made tremendous contributions in the field of protein purification by developing a technique called starch gel electrophoresis. Oliver, who I met on a number of occasions, had the whimsical British charm of the dreamer.

Gordon Dixon had done his doctorate with Oliver and was present when he first tested his theories on gel electrophoresis. Oliver built a crude device to test out his theories on how ionic effects on different proteins would affect their movement through an electrically charged semi-solid environment – his theory being that, depending on the charge of the individual molecule, they should move at different speeds through a gel, thus facilitating their separation. Oliver set it up, applied the test materials, and then suggested to Gordon that they go fishing, saying that it would be hours before they could determine if his ideas had worked. What was the point of sitting around and waiting? If I were the one testing a theory that could change the way in which research on macromolecules would be conducted for the next fifty years, I would not have had the detachment to leave my nascent discovery alone while I went off fishing.

Oliver was inspiring and entertaining to talk to. He bore a striking resemblance to Alec Guinness and had the same slightly crooked smile. And Gordon was tremendously helpful to me when I was beginning my research.

I had never taught before, and I was assigned the responsibility of teaching a course in food microbiology. I would have dearly loved to teach immunology, but those courses were the property of a professor named Cecil Duff. Dr. Duff was getting on and was still using the same lecture notes he had used when he taught me six years earlier. Even then those notes had been out of date, as the field was growing exponentially. I remember their being yellow with age and a bit frayed around the edges. I vowed to myself that I would never be that kind of teacher and that I would update my lecture notes every year, which I did in all the years I taught.

I wasn't crazy about teaching food microbiology. Lectures on food poisoning could be made interesting. But the course was supposed to cover food processing and standards for students taking dietetics and food agriculture, so I had to produce lectures on industrial food processes involving bacteria as well as microbial testing for food regulations. And I was expected to take students on field trips to abattoirs, bakeries, dairies, and breweries. Sadly there were no local wineries in those days – although there were breweries, and those were the highlights of our field trips.

I will never forget the awful smell and the shock of seeing animals slaughtered in the abattoirs. Not surprisingly, someone always fainted during those field trips. I think it was in this course that I developed an antipathy for field trips and the meat industry. Frequently our hosts were ill-prepared and would keep us standing around in cold and damp surroundings expounding needlessly on details no one was interested in. I also gave up eating red meat after our visit to an abattoir.

I spent a couple of years getting adjusted to my new life as a university professor. In 1961, I became pregnant with my second child. Howard and I had been having our difficulties and, in hindsight, my pregnancy was not a wise decision, although I'm very happy I have Ben. I think our problems stemmed from Howard's unhappiness about his own career path. He was drinking a lot and he became mean and miserable when he did. The communal house we lived in made it very easy to slack off and drink too much. There was always someone ready to sit down and have a beer or a gin and tonic.

Howard was not happy with his graduate program. He had tried to write short stories and that had been unsuccessful. We had convinced ourselves that another child would provide stability. It would give Howard the discipline to complete his degree. We were wrong.

My work was going well and Nicky continued to be a delight. He seemed to thrive in the rather bohemian environment in which we lived, as the only child in a house full of adults who, by and large, spoiled him. My mother had retired and my sister had returned to Vancouver from London, Ontario with two children, two little boys. My mother wanted something to do with herself so we made an arrangement that Pamela and I would hire a nanny for all the children, who would spend their days at my parents' house, supervised by my mother while the nanny helped. Thus, Nicky spent his days with his grandparents and two cousins, one older and one younger than he. Pamela was able to go back to teaching so she and her husband could buy a house.

Things between Howard and myself deteriorated. When I was about six months pregnant we had a terrible fight and I decided to end the marriage. But I didn't think it wise to do it then and there. I would wait for the appropriate time to separate from him with the least fallout for Nicky. I would wait until after the baby was born.

I had been having a lot of uterine contractions, but my doctor said they were just something called Braxton-Hicks contractions and not to worry. One morning in mid-July these contractions seemed more pronounced than usual. I was in my lab but called my doctor, who told me to come by. He informed me that I was in labor, and told me to get to the hospital where they would try to slow things down. I was just over six months pregnant and had not told anyone in the department that I was pregnant. Loose shifts were in style, along with muumuus, and I had only gained nine pounds; so, under a lab coat, my condition was hardly apparent. It was embarrassing to let people know that I was about to produce a baby. I had felt uncomfortable sharing this information with any of my male colleagues. At that time, pregnancy was not something you discussed easily with your male boss. You kept your pregnancy hidden for as long as possible, as though it was something shameful.

By keeping me as immobile as possible and heavily drugged, they were able to delay Ben's birth for eight days. Then my waters broke, and he was born several hours later on July 23, 1961, ten weeks early. He weighed just over a kilo. I was told he would likely die.

In 1961, the approach was to take no special measures to save very premature babies for a couple of days. His risk of cerebral palsy and other perinatal disorders was very high were he to survive. I spent the next few days in a state of numbness, afraid to hope. I would go down to the preemie nursery and stand looking in at this tiny creature in the incubator, with legs no larger than my index finger and the face of an old man.

I suggested to Howard we name him after his father, who was called Ben, so our son became Benjamin. A few months later, Howard's father asked me why we had called him Benjamin. I thought he was joking so I gave him a look and said, "We named him after you, of course."

"But my name's Bernard," he said. Oops. But I preferred the name Benjamin, so that was okay.

Benjamin had a strong will to live. The hospital pediatricians decided to put a tube into his stomach on his second day of life because he showed no signs of brain injury or breathing difficulties. I remember he cried as soon as he was born, before the umbilical cord was cut, so he had no oxygen starvation during birth. I started to gain hope that he would be all right.

He remained in the preemie unit for six weeks. After I went home from the hospital, I'd visit there every day and stand outside the preemie ward, looking at Ben in his incubator with tears running down my face. He was so tiny and lonely. I just wanted to hold him. But he gained weight and suffered no setbacks. He came home at the end of September. He had to be fed every three hours when he came home, since he weighed only two kilos by then. The academic year was in full swing, so I was extremely busy and very short of sleep. But Ben thrived and gained weight and was soon sleeping through the night and eating like a lion.

The hospital at that time was participating in a longitudinal developmental study of premature infants in comparison to full term babies. I was asked to participate and willingly signed up. So Benjamin's progress was assessed thoroughly at six-month intervals until he was six and yearly thereafter until he was twelve. (The findings were a great help, as I will discuss later.)

My research was beginning to take shape. I had applied to the Medical Research Council for grant money to support a research proposal and been successful. That enabled me to support two graduate students and start my own research program. And I had figured out how to give lectures that interested students and kept them engaged. I was young, strong, and confident. Life in the communal house was exciting and full of surprises. But my marriage was over.

I knew Howard was as unhappy as I was. We separated the following summer. He stayed on in the communal house and I found an apartment not too far away for the children and myself. Ben was almost a year old by then and doing fine. Nicky would be five in November and was ready for kindergarten. I sent him to a full-day kindergarten and found a day nanny for Ben. Nicky was picked up and delivered by a school bus, so my life had a rhythm to it that I could live with. I spent my days at the university doing research and teaching. I had to be home by 5:30 p.m. to spend the next two hours with the children, feeding and bathing them and then reading to Nicky, something we both enjoyed. Any paper work I had to do was done after the children were asleep. Howard was in library school and took the children out every weekend. We were on quite good terms after we separated and he seemed to think we'd get back together again. I knew differently.

I wasn't lonely. I had good friends in our department, which was growing. A couple of younger people had joined the faculty in microbiology so I didn't feel so isolated amongst the aging faculty members. It did not bother me that there was no man in my life.

One thing that astounded me during those years I was on my own was the number of my married male colleagues who seemed to think I was dying to have an affair. In many instances I knew and liked their wives. I was always surprised by these approaches. These were men I viewed as colleagues; their sexuality never entered my mind, and when it was forced on me by their actions I always felt profoundly disappointed in them. To find that someone I regarded as a friend wanted me to go to bed with him made me wonder about the state of his marriage. Was his wife also involved in dalliances? I was never afraid of those men, because they had no power to hurt me and we were equals, unlike the situation I had found myself in with David Long.

In the apartment I was renting, the people who lived upstairs had a large schnauzer bitch that went into heat the first summer I was there. In those days, most dogs roamed free, people just letting them out their front doors. An incredible variety of mutts, from minis to maxis, appeared from nowhere to sit around, staring at the house looking mournful until hunger drove them home. Although hormones alone inflamed those poor dogs, they reminded me of the way some men will pursue single women.

Going to scientific conferences in those years also underlined how male-dominated the sciences were. Female attendance in those days was very low; a few female graduate students and faculty would attend but the vast majority of attendees were men. The Federation of Biological Sciences meetings and others like them were attended by close to forty thousand people. In the evenings, unattached men would roam the streets in packs of a dozen or so, looking for action. Women were scarce, so I imagine most of these men ended up drinking too much beer together.

Many of these men regarded this time away from wife and children as party time. Science came second. There were any number of evening social events at these meetings, sponsored by pharmaceutical companies. These cocktail events were attended by men looking to ingratiate themselves with opinion leaders or who wanted a leg up in their careers, or to pick up a woman if they got lucky.

Striking up a conversation over a drink with strangers would almost inevitably involve the initial question about whose lab I worked in. My response that I had my own lab never failed to surprise people. I had stepped out of the traditional female role in the sciences.

I didn't live like a nun. I did date eligible men while I was single. But I usually knew quite soon that those relationships were not going to be lasting. And I was fine with that. I had children and a significant career. I was not languishing in the hopes of finding a life partner.

During the sixties, I gained more confidence in my scientific capabilities. I continued to be interested in understanding what was happening at the molecular level when antibodies and antigens bound to each other. But I decided I didn't want to work on the botulinus toxin any more.

There were two reasons for this decision. In order to carry out the kind of research I wanted to do, it was necessary to work with a pure antigen whose structure was understood. To that end, I had put a lot of effort into purifying the botulinus toxin, and although I had had some success in this endeavor, I realized that I would be able to do much better immunological research if I were to work with pure proteins that were readily available and easy to work with. I wanted to define the interaction between antibodies and the antigens with which they reacted. It was now possible to define precisely what parts of a protein molecule interacted with antibodies, if one worked with proteins that were fully understood in terms of their composition and three-dimensional structure.

The other reason I had for abandoning work on the toxin was that I had a visit from two scientists who were based in Fort Detrick, Maryland, known to be the location where US army work on biological warfare was being done. They had seen a paper I had published and were interested in collaborating. That scared me. I decided I wanted no part in research that could advance the creation of biological weapons.

Claude Dolman was upset and angry that I was going to abandon work on the toxin. We became much less friendly. But I felt good about my decision.

This decision entailed a switch in my research direction, and I was uncertain about how such a change would affect my ability to get research support through granting agencies. I had been successful in the past applying for federal grant money to support the costs of my research on the toxin. Grant agencies look for success in ongoing research programs to keep the grant money flowing. Grants were most commonly awarded for three years, so changing your research proposal in midstream did not usually go down well. But I was honest about my reasons for the change of direction and was successful. I explained my wish to change the antigen I wanted to work on and the grants committee approved. I have never regretted the decision.

Claude Dolman retired soon after this. At about this time, the oldest cadre of professors in the department also retired. A new department head was appointed. New faculty members were hired and, to my delight, I was given the opportunity to teach immunology and freed from food microbiology.

At that time, David Suzuki and I became friends. He had a vibrant lab of graduate students and, like a number of us, was separated from his first spouse. He was high-energy and a great organizer of group activities. He would orchestrate camping trips with our kids and some of our graduate students to various destinations during the long summer break. We'd form a caravan and hop onto Gulf Island ferries or drive into the interior of British Columbia.

Long Beach on Vancouver Island was one of our favorite spots. In those days, there was only a logging road from Parksville across Vancouver Island to Tofino on the west side of the island, and the beach was frequently completely deserted when we arrived. We'd drive our cars onto the beach, set up camp, and feast on the giant mussels that were so abundant then. We went scuba diving and smoked pot.

The university in the sixties was awash with recreational drugs. I was cautious. I had seen some casualties among graduate students from taking LSD. I valued my functioning brain and wanted to take care of it. So I restricted my intake to the occasional joint.

Once university lectures were finished for the long summer break, our hours were very flexible. We could take long weekends without depending on statutory holidays. I never felt we were shirking because we worked very long hours when we were in the lab. Because of the children and my need to be home with them, I often met with students or held seminars at night in my home. Otherwise, I spent evenings on paper work, marking exams, writing papers or grant applications, or going over lab results.

The sixties were a very exciting time to be at a university. The Vietnam War was causing chaos in the United States and Canadian universities were gaining from the influx of bright young Americans crossing the border because of the draft. But we watched in horror the assassination of two Kennedys as well as Martin Luther King. It was a time of politicization, student protest, and radicalization – although much more so in the humanities than in the sciences.

Another very exciting thing was happening in the sixties in the form of the feminist movement. Women were waking up. We had the pill. Campuses were stirring with a new awareness of inequities between men and women. Women were becoming vocal. I joined a women's reading group, where we read and discussed feminist literature. This was all very new to me. I had accepted the world as a male-dominated. But I had never adhered to the limitations that this domination placed on women.

The university became aware of the fact that very few women sat on important university committees, like the senior appointments and promotions committee. There were very few women faculty. I ended up being invited to sit on more committees than I could handle. I didn't like administration and regarded it as a waste of time that could be used for research.

My life was very full. Science and my children kept me focused. As I had hoped, Howard seemed happier after we separated. He had decided to go to library school rather than finish his degree in history. After graduating, he got a job at the University of Victoria, met a woman he later married, and got on with his life. I was happy for him.

By the late sixties, my career was taking off. I started getting invitations to present my results at international meetings and was asked to sit on grants panels at the Medical Research Council in Ottawa, the chief source of research funding for medical research in Canada. My work on the nature of antigenic determinants (the actual molecular structure of the individual amino acids that went into forming the three-dimensional shapes that antibodies recognized) was being noticed internationally, and quality publications brought in grant money. I started getting invited to attend and speak at select meetings like Gordon Research Conferences and Keystone events.

Immunology as a scientific discipline became popular in those years, and suddenly the kind of work I was doing was in style. Papers from my lab started being accepted in significant journals like the *Journal of Immunology and Biochemistry* and the *Journal of Experimental Medicine*. Grants are awarded to individual researchers by a committee of peers, on the basis of the quality of the research being done. Publication in prestigious journals always helped the ratings any grant proposal received. It also helped build the recognition our lab was beginning to receive. As a result, I was promoted to Associate Professor and received a big pay increase.

I decided to buy a house. I felt the children needed a garden and a feeling of permanence. I found a place in Kitsilano, not far from where I was living. It was a small, brown shingle house, built somewhere between 1910 and 1920 with two small bedrooms downstairs and an attic with two more bedrooms. The living room, dining room, and kitchen occupied one side of the house. Both the living room and dining room, which were really almost one room, had lovely dark paneling and the dining area had a built-in art deco sideboard and cupboards. There was a small self-contained apartment in the basement. It was perfect for me.

I was very angry when I discovered at my bank that they would not give mortgages to women. I had a tenure-track teaching job at the university, yet I could have gotten a mortgage only if it was in the name of my husband, who was at that time registered as a student in library school! In the end I got a mortgage, properly papered, from my father, who knew he was making a good investment.

The basement apartment made it possible for me to hire a live-in housekeeper. I had a brief but unsatisfying experience with a young woman in my first hire, who turned out to really be looking for a mother to hang out with in the evening, which didn't suit me at all. Then I found a grandmotherly Scottish lady called Mrs. Gibbs who suited us well. She had her own social life, so when I got home from the university I had the kids to myself, which was what I wanted.

Benjamin went into grade one in 1967. As I had done with Nicky when he started school, I sat with him several days a week and reviewed how he was doing with reading and arithmetic. He had no difficulty with numbers and seemed quick, but I noticed a big difference between him and his brother in his ability to read.

Nicky had read quickly and fluently. But with Benjamin, I'd feel his little body go tense as he sat beside me and struggled. Fortunately, as a result of being part of the longitudinal study of premature children born at Vancouver Hospital, he was being tested regularly. He was diagnosed with dyslexia when he was halfway through grade one.

I had never heard the term before, but quickly learned about it. The study he was part of was already showing that about 70 per cent of premature male children were either moderately or severely dyslexic. Benjamin was diagnosed as being moderately dyslexic.

The therapists I consulted did not feel his condition was severe enough to take him out of regular school. I was advised to put him in a private school where there would be smaller classes, so he could get the individual attention he needed. In June of 1968 I knew his reading skills were well below where they should have been. I enrolled him at St. George's, a good private school in the city, and decided that he should repeat grade one there. It wouldn't be humiliating for him, because he'd be meeting different children and he'd get the grounding to give him confidence going forward. But when I talked to the headmaster of the grade school at St. George's he overrode my wishes, saying that they'd make sure Ben got the attention he needed and that they would bring him along. They would put him in grade two. Foolishly, I let the man talk me into that, something I will always regret. The following year would be the worst of his young life.

I was thirty-four. I was happy with my circumstances. I loved the research and I really enjoyed teaching. My children were a joy to me. I had come to the conclusion that it was probable that I would never find a partner with whom I wanted to share my life. The happily-ever-after scenario following marriage was highly unlikely. I had reached a level of maturity where I realized that if you had reservations about a potential partner and went into a relationship thinking that they would change, you were deluding yourself. People rarely change. So I made a mental checklist of the things I wanted in a relationship and decided I wouldn't compromise. The things I had on my list included common interests, a good sense of humor, respect for each other's goals, moral and emotional support for what I chose to do, integrity, and honesty. If a potential partner came up short on anything on my list, that would be that.

CHAPTER SEVEN

Earning My Stripes

People who have never worked in a lab probably imagine that the life of a scientist is a lonely one, stuck away doing experiments on one's own, never socializing. While I don't deny that this might be true in some instances, nothing could be further from the truth in most of the labs I am familiar with. A rich parade of personalities passed through my lab over the years. I made lasting friendships. I learned to accommodate people of widely differing personalities. I learned to mediate and negotiate. Lab life was a vital social activity that shifted and developed over the years. I saw love affairs flourish or fail. There were also occasions when deep enmities developed, never to be overcome.

But most of the time it was a lot of fun. Most of the people in my lab were transient, graduate students and post-doctoral fellows. Graduate students usually stayed between four and six years and post-docs between two and four, long enough to develop meaningful relationships. There were also all the people in other labs that made up the fabric of the social structure I inhabited for most of my career. We worked and played hard; and always, overlaying everything else, there was the relentless drive for scientific truths. Most of us with labs spend more time with this extended family than with our own family.

I loved the lab life. Students keep you young, and their excitement about designing their own experiments was a lot of fun, especially when they came up with ideas I hadn't thought of.

Unlike a number of faculty I knew, I continued to do experiments myself, mainly because I actually loved the hands-on "doing" of experiments. Some people prefer to supervise and administer. Even when my lab became so big I didn't have the time to follow up a project of my own, I enjoyed acting as a technical assistant to a student or post-doc who was running a big experiment.

I still kept my strict regimen of being home by six to spend time with my kids. Now that I had a live-in housekeeper my evenings were certainly freed up, but I always made sure I was there for their dinner and bedtime rituals. I saved paper work for after they were in bed.

By the early spring of 1969 I knew Benjamin was not having a good year at St. George's. Far from the encouragement and coaching promised when I met with the headmaster, the atmosphere in that grade two class was highly competitive and Benjamin was made to feel stupid much of the time, which he wasn't. I didn't understand this until well into the spring term. Benjamin was a very brave little boy and stoic in many ways. He'd go off cheerfully enough in the morning and say little about what was happening in school when he got home. I monitored his reading and math skills and saw his continued frustration with reading, although he was doing better. With math he showed quite remarkable ability. According to specialists I spoke with, the dyslexic sees the spaces around and between the letters on a page rather than the letters. Math is probably easier, since the number variations are much more limited so the space combinations are fewer.

It was in the spring of 1969 that Ben started developing "stomach aches" so as to avoid going to school. I knew by then that he was desperately unhappy. He developed a rash. There were only a couple of months left in the spring term, so I sat down with him and promised him I'd figure something better out for next year. I wouldn't send him back to St. George's. He told me later how he had been bullied and treated like an outcast by his schoolmates. I still get a sick feeling in my gut when I think about what I put him through.

The winter of 1968–9 was the most severe one I can remember in Vancouver. It started the week of Christmas. I remember I planned on having five people to dinner on New Year's Eve when there was a blizzard. I was really gratified when everyone I had invited managed to get to my house and that everyone ended up staying overnight as the snow piled up.

Early in the new year I had a discussion with a fellow faculty member in the microbiology department, Tony Warren, who was interested in creating a new kind of single introductory omnibus course in life sciences that would include microbiology, botany, and zoology. It would be a science equivalent to the Arts I program that had become popular in the arts faculty, which combined English, history, and philosophy. Tony suggested we talk to someone who was teaching in Arts I, and asked if I knew Ed Levy. I said yes, I had met him once. Tony was having lunch with him in a couple of days and asked me to join him.

In the summer of 1968, David Suzuki had introduced me to Ed. He had been an exchange student in Moscow while completing his research for his doctoral dissertation. His field was philosophy of science, and his dissertation was on interpretations of quantum theory. David had said to Ed that he was fed up with science's role in the Vietnam War, and that he was thinking of moving his lab to the USSR. He had been invited to spend several weeks in Moscow. Ed had kindly said he would collect together a few people in the university who had spent time in the Soviet Union to provide a picture of what academic life was like there. David asked me if I'd be interested in going to hear what was being said. I agreed to go, and met Ed for the first time. The evening was interesting and I went away from it with a positive impression of Ed, who had seemed friendly and understated, careful about passing judgment. I thought him handsome. He was tall, slim, and loose-limbed-looking with brown eyes and a warm smile. There was an air about him that told me he was at ease with himself.

The day eight months later when we had arranged to meet Ed at the faculty club, Tony phoned me saying he would be delayed until 12:30, so would I meet with Ed at noon as arranged with Tony's apologies, saying he would join us later. I didn't like that idea. I knew very little about the Arts I program and hadn't spent a lot of time thinking about what a science equivalent would look like, so I said I thought we should cancel the lunch and reschedule when Tony was free. He said fine and asked me to make the call because he had to go into a lecture. I tried Ed's university line repeatedly as noon approached but he wasn't picking up. So shortly before noon, I left a message for Tony saying I'd see him at 12:30, put on my boots and coat and headed across campus to the faculty club. I am basically very shy and really didn't like the idea of this meeting.

The faculty club had an upper foyer by the front door, then a wide half flight of stairs down to a landing with couches and chairs around a big open fireplace. The staircase continued down to the formal big dining room looking out over the ocean. Another stairway led to the ground level where there were meeting rooms and a large cafeteria.

Ed was standing in front of the fire when I got there. I approached him, feeling foolish, and identified myself and gave Tony's apologies. He said he did remember me. He seemed very pleasant and I felt at ease. We sat down in front of the fire to wait for Tony.

I don't remember what we talked about but I remember being surprised when I looked up and saw Tony coming up the stairs from the downstairs cafeteria. I was confused and looked at my watch. It was almost 1:30. Tony had arrived at 12:30 as promised but had assumed we'd be in the cafeteria and hadn't noticed us as he went down to the ground floor. He had to get back for another lecture. Ed and I proceeded to have a late lunch. Again I don't remember what we talked about, but it wasn't exclusively about Arts I.

We became friends. We attended a couple of university functions together. I remember a lecture by a scholar from China who described the Cultural Revolution in glowing colors. I liked Ed very much. He called me one Saturday and asked if I'd like to go ice skating. Beaver Lake in Stanley Park was frozen over and was open for skating – something that happens once every thirty years or so. I was torn. I am a lousy ice skater. I never got the hang of it as a child and therefore didn't like it. But I liked Ed and was curious. Nicky was going out with friends that afternoon but Ben would be with us. I wondered idly if this was a "date." I had heard that Ed was dating a graduate student in the zoology department, and our friendship hadn't gotten to the stage where I could ask him.

I was in my living room at the window when he arrived in his Volvo. There was someone with long dark hair in the passenger seat. Oh, I thought, that must be his girlfriend, and felt disappointed. Oh well, I thought, at least that's clear. Ed came to the door, saying he'd found out that we could rent skates down at the park. When we got to the car I saw that the person in the front seat was a man with very long hair. He was a friend of Ed's who had dropped in to see Ed and was getting a ride downtown. My day got brighter.

It was a beautiful day, sunny and crisp. Ed had grown up in New Orleans and was even worse on skates than I was. But it was such a novelty to be skating outside in Vancouver that the quality of skating was irrelevant. Ben had a good time and skated rings around us. There was a lot of snow on Beaver Lake and the ice was lumpy and uneven, so even good skaters couldn't exactly show off.

Ed lived on an old boat called the Tequila, which was tied up in Coal Harbour, just outside the entrance to the park. After skating we walked over to the boat, which was warm and cozy and had hot chocolate.

I had a serious talk with myself when I got home. I knew I really liked Ed, more than any man I'd met in years. I knew more about him now. He was divorced with no children. He'd done three years in the American navy as a payback for taking a military scholarship to go to university, and he was almost five years younger than I. And I was a single mother with two young children. Why would he want to encumber himself with so much baggage? I told myself another thing. No matter where our friendship might lead, I knew I wanted him in my life as a friend.

Our friendship continued. We went to university talks and events but didn't really "date." We obviously enjoyed each other's company. A couple of weeks later, he called and asked if I'd help him one Sunday dumping some dye into the mouth of the Fraser River. Someone he knew in the geography department was doing a project following tidal currents in that spot. I said sure. Ben and Nicky were with their father that day, so Ed and I were alone for the trip. I remember it being cold and windy and the boat churned in the current when we were dumping the dye. We both got quite seasick and limped back to harbor feeling queasy. One thing about seasickness is that when the wallowing motion stops, so does the seasickness, so as soon as we got back to port, we both felt better.

I realized Ed had something on his mind. We sat over weak tea and talked, and he got around to our relationship in an indirect way. I didn't contribute very much to the discussion. I sat and listened. Ed haltingly seemed to be saying that he didn't want to lead me on. It was kind of the conversation I had anticipated, so I said I understood, that I enjoyed his company and liked to spend time with him but I didn't expect things to go any further. But I was bitterly disappointed. By this time I really wanted a deeper relationship. I drove home thinking that this afternoon would go down as one of the most unpleasant I'd had in a long time – getting dumped and made seasick at the same time.

Oh well, I thought, I guess we'll just be friends.

He phoned me that night and said he'd been foolish. He wanted our relationship to go forward. I slept well that night.

It turned out that Ed had been made to feel guilty by a woman he had dated casually a couple of months before he and I became friends. She had admonished him, thinking, I suppose, that he would only be interested in a casual relationship with me also. She had brought up the subject of my having children and accused him of being someone who wouldn't want take on those kind of responsibilities. He had felt guilty; hence the conversation with me.

We started dating. All the boxes by my mental checklist of "must haves" had ticks in them. We fell in love, and I realized that I had never been in love before. In June he asked me to go to New Orleans with him to meet his parents. We then traveled to North Carolina to meet his brother, Robert, who was working that summer in an Upward Bound program with disadvantaged black kids. And we got married there in the arboretum of the University of North Carolina at Chapel Hill in the presence of Robert, a friend of his, and two black kids from the Upward Bound Program who had happened to be walking by as we were walking to the arboretum. Ed was Jewish and I was Anglican and we were married by a Presbyterian minister who had lost his church position because he strongly opposed the war in Vietnam.

Ed moved in with the boys and me when we got back. My mother had been staying with them while I was gone. I phoned and told her I was married and she immediately told Nicky and Ben. They both accepted Ed with willingness and curiosity. The transition to having a man in our house never surfaced as a problem, although it became apparent a few years later that Nicky had held some initial resentment because he had felt like he was the man of the house. I had been insensitive to that feeling and regretted not having handled things differently. But Ed was such a kind "stepfather" that there was never any strife. His role was always supportive. Many years later, Ben changed his last name to Levy because he said he realized that Ed had been more of a father to him than his own dad.

Ed had a dog named Charlie who also became a member of the family.

We still had to figure out where Ben would go to school. His year at St. George's had been a disaster. He had become antisocial and defensive with other children. My feeling was that he was such an emotional basket case after his year at St. George's that he needed time to socialize with a peer group without too much learning pressure.

An experimental school called The New School had been started by some faculty people from UBC a few years earlier. It was an unstructured establishment with kids expected to learn at their own speed. There were a lot of field trips and a lot of teachers ready to help any student who wanted to learn. It seemed an ideal option for Ben, so I enrolled him there. He became a changed child. He was happier and he had friends. I had no concerns about his being averse to working on his reading. He had been so damaged emotionally by his experience that he needed to heal before resuming any studies. At some point he would decide he wanted to learn. He did, two years later, when he told us he was tired of being "dumb" and wanted to return to regular schooling. He said he wanted to learn, but whenever he had a choice between playing and studying, he always played.

During the summer we all visited a commune at Storm Bay in the Sechelt Inlet. Ed had spent the previous summer there on his boat, completing his thesis, and had made many friends there. The inhabitants of the commune were typical of many such social groupings in the sixties and seventies – war resisters from the United States, disaffected university dropouts who wanted to go back to the land, and common or garden-variety hippies. A lot of marijuana was smoked and people went around naked, weather permitting.

When Mrs. Gibbs, my housekeeper, decided in the fall that it was time for her to move on, we invited a couple Ed knew well from Storm Bay to come and live with us and be housekeepers. They had wanted to come to the city, and this was an opportunity for them. Carol and Merlin were Americans, she from a wealthy Los Angeles family and he a prototypical back-to-nature hippy and a falconer.

We wanted to have a child and I became pregnant that fall. We decided to look for a larger house. We wanted to stay in Kitsilano but found that most houses the size of the one we were looking for had long been subdivided into suites. At that time Kitsilano was a neighborhood of renters. We did find a house in Kerrisdale, the neighborhood I grew up in, south of Kitsilano, composed mainly of single-family dwellings and about the same distance as Kitsilano from the university. The house was located right behind a primary and secondary school. Two developers had bought the house, refitted the kitchen and bathrooms, and painted it. All other aspects were left in their original 1912 state, with lovely old beams in the living and dining room. There were four bedrooms upstairs and a large room in the basement readily convertible to a self-contained suite.

We moved into the house in January of 1970.

CHAPTER EIGHT

Seeing the Light

Our daughter Jennifer was born on July 24, a day after Ben's birthday. My waters had broken three weeks early and she was a small baby, weighing just over two kilos. She was nowhere near as tiny as Ben had been. But she was a beautiful, healthy little girl and we got to take her home when she reached two and a half kilos.

In the month before she was born, Ed and I had taken the Tequila to the Gulf Islands, a cluster of beautiful small islands between Vancouver and southern Vancouver Island. We would anchor the boat and canoe in to isolated beaches to explore, swim, or get clams. Sometimes we fished for ling and rock cod. Nicky had gone to a lacrosse camp and Ben had come with us. We left the boat in Nanaimo and came back to Vancouver about a week before Jennifer was born. When she was three weeks old, we took the ferry over to Vancouver Island and continued our exploration of the islands on the Tequila. Although the southern Gulf Islands are gloriously beautiful, our hearts yearned for the more isolated islands to the north, where we had been the previous summer. But with Jennifer so little, we thought it wise to stay nearer civilization that summer whenever we took holiday time off.

We had been aware for a while that the freedom we'd enjoyed in exploring the BC coast was probably not going to last, as the province became more populated and more and more people bought waterfront properties. The road to Long Beach was going to be paved, so the isolation of that beautiful spot would soon be gone. Long Beach is on the west coast of Vancouver Island near the town of Ucluelet, which was a small fishing village. Its miles of unspoiled beaches were a haven for adventurous campers and divers. To get there, a hazardous unpaved logging road ran across Vancouver Island from Parksville. After the road was paved, Ucluelet became a thriving tourist destination and the character of the beach itself was completely transformed.

We decided to look for a remote property that we could keep in its natural state in perpetuity. We fell in love with a spot we had found the previous year on Sonora Island, one of a cluster of islands between Vancouver Island north of Campbell River and the mainland. The property was in Owen Bay and stretched between the ocean and a small lake called Hyacinth Lake. Between the lake and the ocean a beautiful stream and waterfall ran through the property.

The land was owned by a retired chiropractor in Dearborn, Michigan, who had inherited it from an uncle. As a young man the chiropractor had helped his uncle homestead the land. We'd contacted the owner the previous year but he said he was in no hurry to sell.

To our surprise, he contacted us the year Jennifer was born and said he was ready to sell. He asked $25,000 for the forty-five-acre piece of land between the ocean and the lake. That was a large sum in those days and we couldn't afford it, so we got together a group of like-minded people and formed a land co-op with eight shares. Each share entitled the owner to a building site, while the rest of the land remained common.

The acquisition of this land not only had a profound effect on the way our children grew up to value nature, it also played a role in framing the direction my research would take.

We spent all our summer holidays at Sonora. The old homestead farmhouse had collapsed but there was a small barn built on the side of an embankment that had more or less withstood the elements, although the floor was uneven and the roof leaked ferociously when it rained. We decided to build a cabin.

There were two companies in Vancouver that were selling "do it yourself" precut components for cedar cabins. The components came with a blueprint and rough-cut tongue-and-groove cedar, windows, and laminated arches that formed the ribs of the cabin. The one we bought was a Gothic arch design.

We had the components shipped to Sonora by barge. Getting the material up to our building site was quite a feat, since the shoreline was a mudflat at low tide and rose steeply for about twenty meters from the water when the tide was in. We successfully hitched a winch to a tree and hauled the bundles of lumber up to the flat land where we wanted to build. Our work party mainly consisted of Ed; Merlin, our falconer housekeeper; Nicky, who was old enough now to be able to hammer nails; and myself, along with Jennifer and Ben. Occasionally other friends and family helped out. Nicky, Ben, and Merlin slept in a tent. The rest of us slept in the barn. Every day we all worked until we dropped. We found an old cedar log in a nearby slough from which we were able to cut about three thousand shakes to cover the roof. We did not get the cabin to a livable state the first year but did get it done the second summer.

There are no roads, electricity, or stores on Sonora, where we were in Owen Bay, so we had to bring in all our food. But crabs were abundant in the bay and lingcod or rock cod were plentiful in those days. At very low tide we'd go out to a spot on the outside of the bay and get abalone. A nearby beach gave us clams, oysters, and mussels, so we ate gourmet dinners cooked on an old wood stove. Vegetables we had to bring with us. Miner's lettuce, sea asparagus, nettles, and wild onions were plentiful on the land. Over the years, we cleared enough land to have a vegetable garden and planted loganberries and raspberries.

The cabin was built in a clearing that had probably been a grazing meadow at the time of the homesteaders. The vegetation around the cabin included an invasive large plant called cow parsnip or hogweed. This plant is quite pretty when it blooms, with large white composite florets the size of cauliflowers. The plant grows to about two meters and has big green leaves that look rather like maple leaves.

I noticed that sometimes the children would develop redness in patches on their bodies. Sometimes, these patches blistered and looked like burns. Occasionally Ed and I would get the same spots, usually on the legs, whereas the kids got them on their bodies too. We figured these patches must be caused by some kind of plant sap, but there seemed to be no pattern of occurrence so we couldn't draw any conclusions.

I consulted Neil Towers, the head of the botany department, who was a plant biochemist. He asked me immediately if there was any cow parsnip around where the kids were playing. I said yes, our cabin was in a field of cow parsnip. I said that can't be the culprit because every year the kids play around in the cow parsnips and we cut it down, but only occasionally do the lesions appear. He explained that the sap of the cow parsnip contained chemicals call psoralens, and only when they are activated by strong sunlight do they cause damage to skin.

Psoralens are a family of chemicals that are activated by specific wavelengths of light. When these psoralen molecules are activated, they cause the formation of a reactive form of oxygen called singlet oxygen. Psoralens in the plant sap are not activated because the green of the leaves blocks the activating wavelength from getting through to the sap. But when psoralens on the skin are exposed to the right amount of light at a specific wavelength, they can be activated and cause burns. Sunlight contains the full spectrum of light wavelengths.

When I asked Neil why the appearance of these burns was so erratic, he gave me a long-suffering smile. "We live in a rain forest," he said. "The sun doesn't shine all the time. The chemical needs a certain concentration of light to become activated."

He went on further, explaining that a different class of light-activated drugs, called porphyrins, were showing some promise in early clinical studies for treating cancers. Apparently these porphyrins, when injected intravenously, accumulated more in cancers than they did in surrounding normal tissue. Activation with light could bring about the destruction of the tumor while sparing normal tissue.

I was fascinated by what sounded like such an elegant approach to treating cancer and immediately started thinking about the possibility of including light-activated drugs in some way in my research.

My research interests had expanded in the 1970s. During this decade, a scientist at Cambridge named Cesar Milstein had developed a technique that would revolutionize immunology, the study of the immune response. He had taken a mouse cancer cell and successfully fused it with a normal mature antibody-producing mouse cell.

The resulting hybrid cell, or hybridoma, had characteristics of both the cancer cell and the antibody-producing cell. Cancer cells are relatively easy to establish in tissue culture in comparison to normal cells. They can readily be "immortalized" so that they can continue to divide in culture indefinitely. These immortalized cells are termed cell lines. Most normal cells cannot be immortalized, and will die after a few divisions in culture. Milstein's hybrid cells, however, exhibited characteristics of the normal parent cell in that they continued to produce the antibody, and of the cancer cell in that they continued to divide indefinitely. Not only did these cells continue to provide an endless source of a desired antibody, but also the antibody thus produced was monoclonal, meaning that it was a single type of protein rather than the mixture of different types of protein naturally occurring in normal human antibodies.

The cancer cell line that Milstein had discovered had the potential to fuse with any antibody-producing cell and could therefore be used to generate a limitless number of hybrid cells called hybridomas, each capable of producing a specific antibody. Milstein made the cell line available to other scientists. He later won a Nobel prize for this contribution to the field of immunology.

As soon as I read about this major advance in antibody technology I wanted to develop hybridoma technology in my lab, and sent for the cell line. Hybridomas were a powerful tool for anyone interested in analytical immunology. The technique of hybridization proved difficult but not impossible.

There was a lot of interest in the seventies in whether or not the immune system could be activated in such a way that immune cells could seek out and destroy cancer cells. I had become very interested in this field.

It was known that many human tumors were potentially immunogenic in that they expressed cell surface markers that were different from the original healthy cell from which they arose. Theoretically, the immune system should be able to respond to cancer cells and differentiate between them and the normal cell type from which they arose. And yet patients failed to develop immune responses to their cancers and eventually succumbed to them. This conundrum still puzzles researchers, but recently the discovery of check-point inhibitors and their role in controlling immune responses has made it possible to expect that, soon, the immune system will be able to be manipulated to become a significant weapon to fight cancer. In the 1970s, we were in the dark.

Back in the seventies, a student in my lab using mouse tumor models had been attempting to create monoclonal antibodies directed against a tumor cell line we were working with. She had succeeded in raising a couple of such antibodies. We were working on the idea of "magic bullets," the term used to describe antibodies linked to cell-killing chemicals such as the chemotherapy drugs that are still used in cancer treatment. The concept behind magic bullets was that the antibody would deliver the drug selectively to tumor cells and thus avoid the general toxic side effects experienced when patients are subjected to chemotherapy.

When I spoke with Neil Towers about the phototoxic effects we had seen with the cow parsnip and he told me about light-sensitive drugs that were being tested experimentally in cancer therapy, I came up with the idea of using the photosensitive chemicals to make magic bullets. According to Neil, these photochemicals were not intrinsically toxic until they were activated by light. This would in principle overcome a major shortcoming of the magic bullet approach. Antibodies conjugated with toxic chemotherapeutic molecules, for example, would pile up in healthy livers or kidneys and cause damage. But if the conjugated agent were inactive when not exposed to light, this damage would not occur. I spoke with my graduate student who was working with the tumor monoclonals, and she liked the idea of creating novel magic bullets with the photosensitive chemical hematoporphyrin derivative (HpD), the molecule that was being tested clinically in cancer patients.

I started a collaboration with Neil Towers. He had access to the chemical and we had the tumor models and the monoclonal antibodies.

To chemically link two compounds together, which is what one has to do in making magic bullets, is not a trivial task. The linkage has to occur under circumstances that will not inactivate either the antibody or the killing agent. Also, the linking process must leave both the specificity of the antibody and the toxicity of the agent intact. Ideally, the site on the molecule where the attachment takes place should be highly specific and reproducible, so that every time a new conjugate is formed it has the same activity as the previous one.

It turned out that HpD was not an ideal photosensitive entity to use to conjugate with antibodies. It was a mixture of compounds that were unstable and unpredictable. We were able to produce conjugates and show that the conjugates were delivered selectively to tumor-bearing mice. Also, light activation caused the tumors to shrink with minimal damage to surrounding normal tissue, so the proof of concept had been a successful one. But the results were far from ideal. Conjugation procedures were not reproducible, and the effects, although significant, were not of as great a magnitude as I would have liked.

By the mid-seventies, the first biotechnology companies burst onto the pharmaceutical scene as a result of the successful cloning of human genes and their expression in bacteria or yeast cells. Both Genentech and Amgen were formed during that time. In Canada, Allelix in Toronto and Biochem Pharma in Montreal were created. Genentech's first product was cloned human growth hormone (hgh), a product that is still used to treat congenital dwarfism. Prior to Genentech's product, hgh was isolated from cadavers and carried the danger of transmitting human viruses along with it, or bovine growth hormone was used, which was slightly different from human hgh and could stimulate an immune response against it. Amgen's first product was the growth factor GM-CFS (granulocyte-macrophage stimulating factor). Similarly, monoclonal antibody technology also formed a platform technology for other biotech companies.

In my lab, we had created some monoclonal antibodies that I thought might have clinical significance as diagnostic or even therapeutic agents. We had a monoclonal that appeared to have some specificity for human lung cancer and one for acute myeloid leukemia cells. I had no desire to start a biotech company, but I liked the idea of interacting with either biotech companies or pharmaceutical companies in developing applications for some of the antibodies we had. Researchers are chronically short of research funding and a royalty or license agreement could bring in welcome dollars. Also, the possibility that something we had created in the lab might actually have practical application was exciting to me, as was the possibility that we could do something that would directly benefit people.

Today, most universities have significant technology transfer offices that can facilitate an academic's interaction with industry. In 1978 at UBC that was not the case. I sought the advice of the person in charge of grants administration, a body through which all research grants applied for by academics were funneled for appropriate administrative signatures. I had been referred to him when I started asking questions about who in administration I might see who could advise me about making contact with people in the pharmaceutical industry.

I came away from the meeting very disappointed and with the distinct impression that I was regarded as nothing but a nuisance by the administrator, who basically told me that people in the engineering departments made their own agreements with industry and that if I wanted to do so, I should just go ahead. It seemed as though the university was not at all interested in technology transfer. The arrangements between faculty members and industry essentially meant that the university was not in the loop at all, even when there were patents involved. At the time the university probably had held less than half a dozen patents on discoveries made by its faculty and staff during its sixty-year history.

At about the same time, three things happened that affected my future. A colleague of mine in the physiology department, John Brown, asked me if I would take one of his post-docs into my lab and teach her how to make monoclonal antibodies, because he wanted to develop some against the gut peptide he was working on. I said I'd be happy to help her.

The second thing that happened was that I was asked to give a seminar in the chemistry department on the conjugation work we were doing. I presented the work that we had done with HpD as well as some early work another post-doc in my lab was doing on boron capture. One of the things I said was that the HpD I was using was unsuitable, but that the approach would be worth pursuing if I had a better photosensitizer. After the lecture, a faculty member in chemistry whom I knew slightly, David Dolphin, came up to me and identified himself as one of the world's leading porphyrin chemists. This sounds like a rather boastful statement, but it was true. David probably is, or was, the world authority on porphyrin synthesis. He was the sole author of four volumes of text on porphyrin chemistry. David said to me, "Just tell me what you want in a photosensitizer and I can make it for you." With that, my collaboration with David Dolphin started. And he had spoken the truth.

The third thing that happened involved my mother, who was in her late seventies. She had to have cataracts removed from her eyes. In those days, cataract surgery was not the twenty-minute procedure it is today. Patients were hospitalized and kept immobile for several days after the procedure.

My mother treasured her vision. She is the only person I have ever known who knitted, read, and watched television simultaneously. So she was looking forward to being cataract free. But a few weeks after her surgery she told me that the vision in her right eye had not fully recovered. She held up a fine pair of embroidery scissors and said to me that when she looked at them they appeared curved at the top, like nail scissors. I told her she should make an appointment to see Stephen Drance, her ophthalmologist, who was then the head of the ophthalmology department at UBC.

I knew Stephen quite well, as we both sat on a government committee in the early seventies that was examining progressive methods for the delivery of health care. My mother saw him and called me with the bad news. She had the wet form of age-related macular degeneration (AMD). She would lose her central vision, and there was no treatment other than a laser procedure that usually worsened your vision and didn't always halt the continued deterioration of the macula. She was stoic, as she always was, but I knew she was devastated. As was I.

I had never heard of macular degeneration at that time. Hoping that she might have missed something, I called Stephen Drance, and after speaking with him realized that my mother had very accurately described their conversation. He explained the condition as blood vessels becoming leaky at the back of the eye, in the macula, the part of the eye that gives us our central vision. This leakage is progressive and eventually results in scar tissue forming, which blocks the passage of images through the macula to the optic nerve. The macula is a small part of ocular taxonomy but is absolutely critical for central vision. Ophthalmologists used lasers to try to burn out the leaky areas in the macula and thus stop further deterioration. Unfortunately, laser burning also destroys the macula, and Stephen did not recommend the procedure. So my mother was destined to lose her vision. Her right eye was worse than her left, but her left eye was also affected.

Although my mother lived until she was ninety-four, she never gave in to her condition, never admitted she was legally blind, and never consented to carry a white cane when she took the long walks she enjoyed. With macular degeneration, one does not lose peripheral vision, so she would catch fleeting images of, say, the moon on a bright night and she would comment on it.

She had dinner with us at least once a week. She confessed at one point that she preferred colored wine glasses so that she could distinguish them and not run the risk of knocking the glass over at dinner. I found an antique, quite beautiful wine glass with a heavy gold leaf rim. That became "Dodo's" glass. The grandchildren had morphed her name, Dorothy, into Dodo, and that's the name all her grandchildren called her.

One evening several years later, when she and I were chatting, she spoke of what she "saw" through her damaged macula. She described the brownish mist that formed the central part of her field of vision and the flashes of image and form she'd see peripherally. Then she said, "The sad thing is that I'll never see your pretty face again." That broke my heart.

CHAPTER NINE

Becoming Part of a Biotech Start-Up

I had never collaborated with an organic chemist before. But David Dolphin proved as good as his word when he said he could make any kind of porphyrin I wanted. Organic chemists work very fast. In contrast, experimental work in biological research labs frequently involves protracted and complicated experimental procedures. Evaluating the potential use of a chemical substance in biological systems necessitates first running the compound through a series of test tube assays. Activities of each compound have to be titrated. Cell cultures have to be grown under controlled conditions. Experiments have to be repeated. Then the substance has to be evaluated in animal models. Each evaluation takes time and resources.

What organic chemists can do, I discovered, is generate dozens of unique compounds in a matter of days. We were inundated with new photosensitizer molecules from David that all had to be tested.

I hired a professional assistant, Anna Richter, to manage the research program involving David's molecules. Anna and her husband Stan had defected from Poland after attending a scientific meeting in Western Europe several years earlier. He was an engineer and she had a PhD in physiology. They had left their two-year-old daughter, Joanna, in Warsaw with Anna's mother and had only succeeded in being reunited with her after four years. Joanna was now nine, almost exactly our daughter Jennifer's age. I wondered if I could have done what Anna and Stan did.

Anna walked into my office one day and said she wanted to work for me. She had heard about our photosensitizer research and had decided that this kind of research was what she wanted to do. She was currently working in a physiology lab at Vancouver Hospital. She turned up just as I was realizing that I really did need someone who could take charge of the photosensitizer work. Anna was an original thinker, organized and highly competent. She took charge of photosensitizer evaluations. I soon trusted her judgment completely. She and I quite often disagreed about the conclusions to be drawn from sometimes puzzling experimental results and would argue over which were the appropriate next steps. Sometimes I was right and sometimes she was. I was always delighted when she could argue me into realizing that her logic was superior to mine, because I always felt I had gained greater insight through her. She and I worked together until we retired.

The collaboration with David Dolphin was exciting. We selected one molecule out of the collection that he had created. It was called benzoporphyrin derivative monoacid ring A, BPD-MA or just BPD for short. If photosensitizer molecules can be described at beautiful, this one was. Porphyrins as a chemical class were so named because most of them are red in color; they were named after porphyry, the red stone used in many classical buildings in Greece and Italy. BPD was green. All the early workbooks reporting on data we gathered on BPD bore the label GS, which stood for "green stuff," until it was labeled BPD-MA. Porphyrins as a class are common molecules, heme being one of the commonest. Heme is the part of the hemoglobin molecule that carries iron around in our blood and the blood of all mammals. Most of the photosensitizer molecules studied at that time were derived from the heme molecule, which is cheap and readily available in pig or cow blood.

BPD-MA had one chemical attribute I needed. It only had one position in the molecule that could be activated to form conjugates with antibodies, thus making the conjugation step more predictable. It was also an extremely powerful photosensitizer, another asset.

My lab was large and we had several distinct areas of research. The photosensitizer evaluation and analysis formed a subgroup along with the hybridoma group. In addition, there were other students and fellows working on immune regulation and cancer immunology.

By this time I had about fifteen people in the lab, including graduate students, technicians, and post-docs. I spent a lot of time writing grant proposals to federal agencies in order to be able to support this level of research. In Canada at that time, the Medical Research Council was only funding 30 per cent of proposals they received, so competition was very stiff. Similar success rates were seen at the National Cancer Institute of Canada, too.

The collaborative work I had undertaken with John Brown, my colleague in the physiology department, had been progressing, and his post-doctoral fellow had successfully produced a number of good hybridomas making monoclonal antibodies directed to his gut peptides and other human hormones.

John and I had become friends. He was British, the son of a coal miner and sensitive about his working class origins. He had gone through university on scholarships. He had a reputation for being a difficult colleague, prone to fits of temper and pugnacity. But I liked him. He had a wicked sense of humor and a big chip on his shoulder, ready to take offense at any presumed slight.

John told me that he and three other colleagues were starting a biotechnology company. I was impressed. He said they needed the hybridoma technology in their company and invited me to become a founder along with them. I said I was pretty busy with my research, but I was interested in being connected to a biotech company because we had some antibodies in our lab that might be useful in diagnostic kits. He said being a founder would not take any of my time. I said okay. It was going to take some months before the company was formally created.

In June of that year, 1980, our son Nicky died in a drug-related accident. He was twenty-two. We had known that Nicky and the friends he had at the time were being reckless and were immersed in the drug culture that had become prevalent in high schools in the seventies. Our attempts at intervention were useless. We felt helpless, watching someone we loved make terrible choices. It was fentanyl that killed him.

To lose a child is the worst thing that can happen to anyone. Your world changes. We coped somehow. Ed, my husband, Ben and Jennifer, our children, and I all suffered in our own ways. Each day for me was a black distance I had to travel. Having responsibilities helps, because you have to carry on even when the chasm of grief that your mind and body have become paralyzes you. I looked for insights I might gain from this experience. Perhaps experiencing grief at that depth gives us a better understanding of other people's grief. I learned that you can't just "get over" something like a death. If you love the person who is dead, your grief will be permanent. It stays with you and eventually gets to be manageable. After a while you can put it in a box and it doesn't come spilling out unbidden. But it doesn't go away. It remains, as it should. I became emotionally absent from everyone around me for many months. Each of us in the family suffered in our own way.

My colleagues at the university were supportive and sensitive. Ed was wonderful, quietly supportive. That summer, two of my British cousin Tony's children came to Canada. They were all grown up now, Tim about twenty-eight and Clarissa twenty, Ben's age. They moved in with us for the summer, which was an important diversion for us.

Towards the end of 1980, John Brown and his associates took the first steps to form the biotech company he'd spoken to me about earlier. There were five founders: John; myself; Jim Miller, a neurophysiologist; Tony Phillips, a neuropsychologist; and Ron McKenzie, an engineer turned businessman. Jim, Tony, and Ron had been classmates at the University of Western Ontario. John knew Jim and Tony. I knew only John. Jim and Tony had been close friends since their college days. Tony was never active in the company other than serving on the board during the first few years.

The company was named Quadra Logic Technologies, a name I have been asked about many times. People ask about where the Quadra comes from. The electoral district most of us founders lived in at that time was called Quadra. There is an island off Campbell River called Quadra Island. History tells us that Captain Quadra was a friend of Captain Vancouver. None of that has anything to do with the name of the company. Ron McKenzie had made a call to Victoria, the capital of the province of British Columbia, in order to determine if the name he had submitted to register the company was acceptable. The name he had submitted was Bio-Logic Technologies. The bureaucrat with whom he spoke told Ron that the name was not acceptable because it was too generic. Ron thought for a moment. Would trio-logic work, he wondered. No, it didn't sound right. Quadra-logic?

That sounded okay and there was some talk of the company raising research animals on Quadra Island, so he went for it.

We had an assortment of monoclonal antibodies to various tumor-associated antigens, John's peptides, and other human hormones. Some of them could form the basis of diagnostic kits. If we were going to form a company around such kits, we needed money to set up a facility to put kits together. We also needed expertise in that area, so we needed money for infrastructure and personnel.

We got some money from a government initiative called IRAP, a well-thought-out government program that provides seed funds to start-up companies. With the IRAP grant and the roughly $50,000 we had each invested we rented our first "office" and hired a technician. It was a small two-room space above a bakery on Forty-First and Dunbar in Vancouver. Forty-First was one of the main routes commuters used to reach UBC, so its location was convenient – we could stop in there on our way to or from the university.

We hired a technician and set up a small lab where we could raise mice for making hybridomas. We produced and stocked monoclonal antibodies to John's peptides as well as an antibody to chorionic gonadotropin, a substance that is elevated in the urine of pregnant women. We had ideas about creating a pregnancy test using this antibody. We were never told we could not breed mice in that facility, and we didn't ask permission. Our office was behind a psychotherapist's office. The therapist practiced primal scream therapy, so it could be quite noisy in the lab. We never told him about the mice.

Not one of the founders of our company had the faintest idea about the business side of a biotech company. Four of us were academics and Ron, the nominal businessperson, was a Zen meditator, it turned out, and given to flights of fancy. Although he had some business experience, he knew nothing about biotechnology. I remember at one point commenting to someone that my reality sphere didn't overlap in any way with Ron's.

As academics, we had all had significant success in our own fields; each of us were recipients of sizeable research grants and therefore thought we could succeed at raising money for the company. I think we all had the attitude that we were smart, therefore we could succeed at whatever we set our minds to. I think of it as academic arrogance. We were wrong.

Jim and Ron assumed the responsibility for making deals and raising money for the company. They were excited about deal making.
I paid very little attention to what they were doing. I was happy
they were out there trying to find companies interested in our antibodies and to raise money that would enable us to start developing
kits of our own. We'd also have to have funds for building our in-
tellectual property. I showed up for meetings when summoned and
presented information when requested, but my heart was not in the
financial side of the company.

Initially, we dealt largely with the Vancouver financial community. Money was flowing on Howe Street but investments were primarily in natural resources, mining, forestry, or oil. The Vancouver
Stock Exchange was active in penny stocks and had a rather questionable reputation. These hardened Howe Street investors in gold
mines, real estate, and oil wells were unfamiliar with even the word
"biotechnology."

I don't remember the details of how we funded the company in
those early days. I do recall one particularly harrowing experience
concerning a one-million-dollar deal that Ron and Jim engineered
with a questionable source of venture capital who had a terrible reputation as being sexually abusive. This investor ended up threatening to close us down unless he got control of the company or was
repaid in full. I could tell Jim was very worried when he told us,
even though he spoke with his usual bravado. We were scheduled
to meet with a representative on the following Monday. I spent the
weekend worrying. I noticed that Jim was pretty hyper that Monday morning and not saying anything. He went out and took a
long walk. I knew he was planning on pulling something off but he
wasn't talking. When we went in to meet with the representative,
Jim passed me a brief note saying "don't worry." The representative
started threatening and demanding repayment. Jim let him go on
for a while and then took a check out of a folder and tossed it across
the table at him. It was a check for a million dollars. I remember the
representative gaping at the check for a moment, in disbelief. But it
was real. Jim had had it certified.

The back-story to the mysterious million dollars was that Jim had
been courting investment from an old hand on Howe Street, a man
called Ray McLean. Ray had been a prospector as a young man and

had been very successful. He was a big bear of a man, probably in his late sixties when Jim met him. The two of them really clicked. Ray had trusted Jim with the million-dollar investment. He became a staunch supporter of the company and served on our board for many years. He and Jim remained close after both of them were no longer associated with QLT, and I believe Jim continued to handle Ray's biotech investments until Ray died.

My contribution to the company at that time was in writing grant applications to both federal and provincial agencies that had resources to fund start-up companies or to fund university/industry collaborations. I quickly learned the differences in applying for money to support applied research as opposed to basic research. With applied research, funders were looking for finite deliverables, a product coming from the research. In basic research investigators are rewarded for making progress rather than coming to a conclusion.

The world of government-supported research, even applied research, where I felt comfortable, was a far cry from talking to bankers and venture capitalists about making investments. I realized I had to get comfortable with this new reality we were entering.

I remember we did a small private offering of shares to "friends and family," a common vehicle for start-up companies. Our investors were mainly university colleagues, many of whom hung onto their shares and did very well.

The company floundered along during the early eighties, investment coming in sporadically from a variety of sources, either from small investments or government-supported research. Jim and Ron were very active in looking for deals. They started traveling widely. Occasionally, Jim would meet someone on a flight whom he took a shine to and he'd hire him on the spot. I remember someone who was briefly with the company whose business card read "galactic explorer." I never knew what this person was hired for. He was a pleasant enough young man who I think had some kind of background in marketing. His first name was Ihor and for a while he was in charge of developing the pregnancy test kit.

Jim and Ron went through a phase of thinking that deals with China could make our fortunes. They took many trips to China to try to broker deals and found a company that expressed an interest in our still-to-be-created pregnancy test kit. A challenge in those days was that China had little cash. One idea they had was to make a profit in the short term for the company by swapping pregnancy kits for fine chemicals from China and marketing them. The short-term cash could support the longer-term development of pharmaceutical products.

I recall that during this time a kind of envoy from the Chinese company was sent over to be at the company to broker trade deals. His name was Henry. We thought he was a spy. He seemed very nice but no one could figure out what he was doing. I remember him playing the violin very badly at a Christmas party.

I discovered that Jim and Ron had expanded the scope of the deal to include all sorts of products, including cashmere sweaters and condoms. I watched bemused from the sidelines as the condoms were tested for quality control by blowing them up like balloons. When they started to explode they were sent back. The fine chemicals showed up in strange egg-shaped plastic containers about the size of ostrich eggs. I never saw any cashmere sweaters.

I have forgotten what exactly happened with the China deals, but I was relieved to see Henry go back to China and the mysterious egg-shaped containers be relegated to storage. Jim had by now given up his faculty position at the university to be a full-time CEO at what we now called QLT. Jim was, and remains, a born entrepreneur, energetic, optimistic, and a persuasive salesman. He was confident that he could raise money and create a successful company. He and Ron did the work to get us a listing on the Vancouver Stock Exchange and we completed an IPO in 1986 and ended up with a few million dollars in the bank.

We moved to more respectable quarters, first to a new research building associated with Vancouver Hospital and then to a fully equipped lab abandoned by another biotech company that had been in the business of cloning rhododendrons and had gone into Chapter Eleven bankruptcy. We inherited a large amount of good equipment along with taking over the lease.

My research on photosensitive drugs had progressed. David Dolphin's molecule, BPD, continued to show enormous promise. We made a rather surprising observation with this molecule. After working hard on the conjugation technology, we were ready to test our conjugates in animal models. Much to our surprise, when we tested the BPD conjugates in tumor-bearing mice, and ran the experiment with all the appropriate controls, we discovered that the drug on its own – one of the control groups we ran – found its way very quickly and efficiently to tumor tissues without needing the antibody component to create the selectivity we were looking for. This surprising observation opened up a development opportunity that was very much easier than developing conjugates.

This observation changed much of my thinking about where the company might go. BPD was a powerful photosensitizer. It could be developed on its own as a therapeutic agent for the treatment of cancer. There had been progress in the field since Neil Towers first told me about light-sensitive drugs and mentioned that HpD, the derivatized hematoporphyrin molecule, was already being tested clinically for treating cancers. Now, photodynamic therapy was being developed as a treatment modality by a legitimate pharmaceutical company, Johnson and Johnson. Their drug, called Photofrin^{TM}, was a slightly cleaned up version of HpD. And our drug, BPD, was better than Photofrin. A business plan could be written around it as a potential therapeutic. I set out to do just that.

CHAPTER TEN

Positioning QLT as a Photodynamic Therapy Company

Before embarking on a business plan, I did some research on photo-dynamic therapy (PDT) as a feasible approach to cancer treatment. At that time I knew very little about PDT. I had been focused on antibody conjugates; now I had to learn the background of PDT.

I discovered a whole new field of biology. There were several different photosensitizer molecules considered promising candidates for PDT. Photofrin or HpD was clearly the lead product, as there were many publications from clinical research labs attesting to good clinical outcomes for a variety of cancers including lung, oral, head and neck, and skin. The publications, however, mainly reported results from small numbers of patients and did not have comparators with other therapies.

Tom Dougherty, a chemist at the Roswell Park Memorial Institute in Buffalo, New York, was the creator of Photofrin. He is known among colleagues as the father of photodynamic therapy. Tom had started a company called Photofrin Medical out of Roswell Park in order to develop Photofrin as a cancer drug at about the same time as we started QLT. His stories about his early years of mixing up batches of Photofrin in a garage are representative of those early years in biotech.

He had sold Photofrin Medical to Johnson and Johnson, who took over production of the drug and were running phase II clinical trials in lung, bladder, and esophageal cancer. Johnson and Johnson renamed their new subsidiary Photomedica.

In our view, Photofrin had inherent problems as a drug. In the first place, it was an unstable mixture of porphyrin polymers. If it was left in solution it underwent chemical changes. Also, it was inconvenient. Patients were injected intravenously with the drug. After injection they had to stay out of strong sunlight for up to six weeks after a single injection because their skin became highly sensitive to light. After two days, during which time the drug accumulated in the cancer tissue, they returned to the hospital, appropriately clad with sunglasses, hats, and long sleeves, for their laser light treatment of the tumor.

Our photosensitizer, BPD, on the other hand, accumulated very fast in cancer tissue and treatment with light could be carried out three hours after injection. The drug also cleared the body very rapidly so that the side effect of photosensitivity lasted less than three days. And it was chemically stable and existed as a fifty-fifty mix of two isomers, which had identical physical properties.

We considered ourselves in a strong position to develop BPD as a second-generation photosensitizer with superior properties to Photofrin. By this time I had learned something about drug development. When we'd started QLT, we thought we could handle the variety of potential products each of us had generated in our labs. Our concept had been that we could create and sell diagnostic kits using our antibodies, or we could sell reagent antibodies. Such a business model was acceptable but was never likely to gain a lot of investor interest, and would require significant infrastructure since growing monoclonal antibodies for commercial use was heavily regulated. Diagnostic companies never attract the kind of investor interest that therapeutic companies do, because profit margins are much lower for diagnostics.

But developing a drug for the treatment of cancer using a new therapeutic approach could well attract investor interest. I realized that to develop just one drug in a small company was a monumental task. Drug development and research were miles apart in terms of cost and time involved. The numbers being used at that time were that the cost of developing a drug that successfully gained marketing approval was in the neighborhood of a hundred million dollars and would take ten years. I believe the number has risen to five hundred million today. The literature contains much higher

numbers, in part because the costs associated with failed development are included, thus doubling the cost of a successful drug.
Of the drugs that have survived a number of years of preclinical
research and gained permission from regulatory authorities to be
tested in humans, only about 25 per cent are actually successful in
gaining marketing approval. Thus, the doubling figure is not out of
line. Of course, I didn't believe the numbers. Unfortunately, they
turned out to be quite accurate.

I proposed to the other founders that I prepare a business plan
for investors, positioning QLT as a PDT company with a promising
second-generation photosensitizer for cancer PDT. I got their buy-in.

As has happened in my life a number of times, a series of events
took place that conjoined to have profound effects on the future
of the company. The first of these was that Johnson and Johnson
suspended the Photofrin clinical trials they were sponsoring. The
reason given was that they had run into manufacturing problems.
While this was true to some extent, the real truth lay deeper. The
deal with Tom Dougherty's company had been championed by
one senior individual at Johnson and Johnson. That person was no
longer with the company, and the remaining decision-makers were
disillusioned with a product that was causing problems and really
not part of their core business. Small companies dealing with pharmaceutical giants have to have an internal champion, I have learned.
Photomedica lost their champion and the project was jettisoned.

John Brown had generated a number of monoclonal antibodies
that reacted with various hormonal peptides. He had one that in
early studies looked like it might act as a growth stimulant in poul-
try. Jim and Ron had been on the road mustering interest in this
monoclonal. One of the companies they had approached was the
animal health division of Cyanamid Canada (the Canadian branch
of American Cyanamid).

At about this time, I was invited to give a seminar in Hamilton,
Ontario, at McMaster University. My talk involved regulation of the
immune response and had nothing to do with the photosensitizer
work I was involved in. At the reception and dinner held after my
talk, I got into a conversation with a woman physician who was
treating breast cancer patients at the Hamilton Regional Cancer
Centre. She told me how disappointed she was that Johnson and

Johnson had stopped their clinical trials with Photofrin and how much benefit her patients were deriving from photodynamic therapy. She was treating women with breast cancer that had recurred on the chest wall around the incision scars from mastectomy. This is a ghastly form of the disease that is very difficult to treat. Most of these women had received radiation therapy on the chest wall after radical mastectomy, which had caused their skin to thin and become frail. Their radiation dosage was probably maxed out so they couldn't be treated again with radiation. Also, the recurrent tumors were frequently resistant to chemotherapy, so they didn't have a lot of options left to them. For the first time I realized that PDT had a real place in the armamentarium against advanced cancer, and that even though Photofrin had its weaknesses it was still benefitting desperate patients.

On the long flight back to Vancouver, I had an idea. We were positioning QLT as a company developing a second-generation photosensitizer. I knew a little about the manufacture of Photofrin and the process had seemed pretty simple. Why couldn't QLT take on the responsibility of manufacturing Photofrin for Canadian physicians? I got excited about this very naïve thought.

When I got back, I called Jim Miller and told him my idea of manufacturing Photofrin for Canadian physicians. Jim, whose positive thinking knew no limits, said, "Make the drug for Canadian doctors? Hell no, we'll buy the whole damn company from Johnson and Johnson."

My response to him was a shocked, "With what?" I knew we had some money in the bank, but not enough to make an offer on a company that was in clinical trials. But I remembered the almost miraculous appearance of the million dollars from Ray McLean and didn't push Jim on the credibility of his remark.

Jim and Ron had become friendly with the then-head of business development at Cyanamid Canada while they were trying the get the attention of the animal health unit about John's monoclonal. It turned out that American Cyanamid had a significant franchise in cancer. In their clinical department at Lederle Laboratories there was one clinician in particular who had been watching Johnson and Johnson lose interest in PDT. His name was Stuart Marcus, and he was championing the idea that Cyanamid should buy Photomedica.

Much later, Stuart told us how hard he had lobbied to get his bosses interested in PDT.

As has so often been my experience in my career, circumstances came together at a critical moment. The Canadian sub of American Cyanamid got the attention of the US people interested in PDT, and we started constructing a deal for a collaboration. We came up with an interesting business model. American Cyanamid was nervous about fully embracing a new and risky technology like PDT. They preferred to keep the financial investment at arm's length from the spending budget for their regular R & D activities. While they had an interest in PDT, they didn't want to bring the technology in-house. They had found us, a small PDT-based company that could be the public face of PDT while they acted as licensees and backers.

We showed up at the right time. The Canadian sub of Cyanamid got credit for finding us. My husband Ed had taken a sabbatical from UBC to do research on mandated science and how the regulatory framework in biotechnology fitted with academic science. He decided to use QLT as the test case. He became involved in some of the due diligence. Ron, Jim, and our corporate lawyers worked all through the fall of 1987 on bringing the deal between American Cyanamid, Johnson and Johnson, Photomedica, and QLT together.

At the same time, we realized we had to raise money if we were going to get serious about drug development. We didn't know how much up-front money we'd be able to get from a deal with American Cyanamid, and drug development was expensive. Jim, myself, and Rashid Aziz, our CFO, went looking for investment bankers to raise funds and do an IPO on the Toronto exchange as well as on NASDAQ, the New York–based exchange that most biotech companies trade on. The fact that we had focused on an interesting and potentially important new therapy and had a credible second-generation product made us an attractive investment opportunity. Nesbitt-Burns was our first banker, and my first experience in dealing with the investment banking community.

Things moved very fast in the fall of 1987. I was very busy at the university, but also at QLT. Nesbitt-Burns wanted to set up a cross-border share offering to get American investors involved. I quickly learned the meaning of "road trips."

Road trips involve investment bankers, their analysts, and company representatives. Most commonly in biotech companies, the presenters from the companies seeking funding include their CEO, CFO, and CSO (chief scientific officer). As the VP of research, and thus CSO, I had to go on road trips to explain the science.

The company representatives are taken around from one potential investor to another in rapid succession, with barely a moment to use the toilet. Potential investors include fund managers, pension funds, and other financial institutions. These road trips can go on for three to four weeks. They are completely exhausting.

Each meeting was usually a thirty-minute presentation. Jim as CEO would introduce the company, followed by my presenting the science and development path. Jim could get carried away with his own enthusiasm and often made statements about the science that were promotional at best and simply not true at worst. I often began my presentations by saying, "What Jim actually meant when he said ..." I commented to Rashid that I felt kind of like the guy at the back of a circus parade carrying the shovel after the elephants have gone by.

At the same time as the planned IPO, the negotiations with Johnson and Johnson were progressing. We met Tom Dougherty, the creator of Photofrin. We knew he was not enamored of the idea of his drug falling into the hands of a small Canadian company that was developing a competing product. I could understand his feelings. But even in those early days he was always courteous and respectful to us. In subsequent years, Tom and I became fast friends and he admitted that, as things turned out, our involvement with his product was the best thing that happened to it.

Our bankers brought in other bankers from Prudential Bache (or Pru-Bache) out of New York to handle the US IPO. Pru-Bache brought in a team of professional marketers to work on our presentations to potential investors. I think both Jim and I felt we were pretty good at making presentations. We both had taught and had made numerous scientific presentations in the past. We learned differently. Our coaches stayed three days. They eviscerated our slides and recreated them with about half the words on them.

I had begun this exercise with a mildly disdainful attitude. I finished it with far better insight into how to present complex information and a great deal of respect for these Madison Avenue professionals. The lessons learned stayed with me. Over the years I have been able to help many small companies improve their presentations when they start out.

We did our Canadian IPO in the late fall of 1987. We raised ten million dollars. Just before Christmas we closed the deal with Cyanamid whereby Cyanamid invested fifteen million dollars in QLT and got marketing rights to Photofrin. We bought Photomedica and agreed with Johnson and Johnson to complete the registration trials for the drug.

The company was on its way.

We had taken on the tremendous responsibility of completing the registration trials (phase III) for Photofrin. None of us at QLT had the slightest knowledge of what that would mean, and how much we had to learn during the next decade. And I was still heavily involved with teaching and research and unsure of just how much time I wanted to spend with investment bankers who were only interested in money.

CHAPTER ELEVEN

Colleagues at American Cyanamid Come to Our Rescue

We had twenty-five million dollars in the bank! We couldn't believe it. We had reinvented QLT in a few short months and things had really come together.

Then reality struck.

Photofrin was a drug that had been through the first two stages of clinical testing. Now it was up to us to make sure the final stage of clinical testing was done appropriately, so that we could get FDA approval.

And that was a problem. We had virtually no in-house expertise. Cyanamid knew this. They wanted the technology but didn't want the project to show up on their books as an R & D expense. They also wanted BPD as a follow-on product; but their approach, which was common in the industry, was to get the first generation, Photofrin, approved, and then use revenue from sales to finance the development of the second generation, BPD.

There was little doubt that PDT had a place in cancer management. In diseases like lung and esophageal cancer, patients suffer terribly when their tumors recur locally and get so large that they block their airways or alimentary tracts so that they have difficulties breathing or swallowing. Bulky local recurrences were treated by putting in stents or with thermal lasers to burn out the tumor tissue. But tumors grow through stents quickly and thermal lasers can cause serious burns.

It is relatively simple to access tumors in the lung or esophagus with fiber optics. PDT might provide a better alternative for palliation of these conditions. Our partners at American Cyanamid (ACY) suggested that the phase III trials should include these two cancers. The phase II trials had already shown that PDT could alleviate the suffering of such patients.

We knew we had to hire a clinical research organization (CRO) to run the Photofrin clinical trials. CROs provide services that take drugs through the various stages of development. Even large pharmaceutical companies use CROs when they embark on large clinical trials, because the staff required can be larger than their own permanent staff. Separate and specialized CROs may be employed to take a drug through toxicology testing and clinical trials. We looked into Canadian-based CROs and were dismayed at the quotes of costs that came back. Our small fortune started to look smaller. In our naïveté we had thought that twenty-five million would take us through to FDA approval.

Our colleagues at ACY came to our rescue. They discreetly offered to act as our CRO for the phase III trials on Photofrin. In our ignorance, we thought we had a special deal. Surely our partner wouldn't charge the same fees as the CROs we'd contacted. We were naïve, first to believe we had enough money to get through the phase III clinical trials, and second to believe we would get a deal from ACY that was significantly better than the CROs' estimates.

We formed a joint development team with ACY on which each function for drug development was represented by a member or members from each company. Functions included toxicology, preclinical, clinical, regulatory, marketing, project management, and devices. The team would meet face to face every quarter to decide on next steps and assess progress. Several ACY VPs attended these meetings, too. We at QLT didn't have appropriate counterparts to the Cyanamid team, so many of us were pinch-hitting, to say the least. I represented preclinical and BPD development. I felt very comfortable in that role. Ed led QLT's participation in the project team.

Now that we had money, we started looking for qualified people. We discovered that Canada was pretty thin on the ground when it came to people experienced in drug development. On the west coast there was no one. In one case, we were fortunate. Alexandra Mancini was the wife of John Mancini, who had recently been hired at UBC to head up the Department of Medicine. They were returning Canadians who had spent the previous few years in the United States. Alex had been employed in the regulatory department of a large pharmaceutical company and had acquired good experience in monitoring phase III trials and preparing documents for submission to the FDA. She was looking for work, and we were looking for her. She became a key player in Photofrin development and our relationship with ACY. She quickly formed bonds with the ACY project leader, Lou Gura, and Stuart Marcus, the ACY clinical representative.

Both our investment banker and ACY wanted QLT board positions, which was understandable. Nesbitt-Burns selected Duff Scott, former head of the Toronto Stock Exchange and then-president of Prudential-Bache Canada. He became chairman of our board and remained in that position until 2004. ACY put forward a former CEO of Cyanamid Canada, Jan Dlouhy, who was then the vice president of research at the head office of ACY. Jan was an elegant figure who was always impeccably dressed and given to dry wit.

At one point after I got to know Jan, I asked him why Cyanamid had structured our deal with them in the way they had. "Wouldn't it have been easier for ACY to simply negotiate with Johnson and Johnson?" I asked. "Cyanamid really doesn't need QLT to get Photofrin trials finished. They could have made a separate deal with us for BPD if they wanted it."

Jan gave me a smile. "It's like a single-storey whorehouse. There's no fucking overhead with this deal," he said.

I found his comment amusing but I didn't really understand its meaning until later. ACY, like all big pharmas, was very conscious of returning high earnings per share to their shareholders every quarter. To get high earnings per share, the company has to keep down R & D expenses, much of which is overhead. ACY could get ownership of two PDT drugs by making a relatively small investment that didn't show up on its books as an ongoing research expense. By arranging with us to have ACY's research group carry out the phase III trials, they turned the Photofrin project into a profit center rather than an expense. We were being charged for their overhead. Even though the deal proved to be an expensive one for us, what we learned was invaluable. ACY willingly shared their knowledge and experience with us.

I look back on that time with great fondness for the people who tolerated our ignorance and taught us the rudiments of how large pharma works. Of course, not every member of ACY's project team was enthralled with the role of tutoring QLTers, but most shared willingly. And we made some significant contributions. We understood PDT and they did not. There is a world of difference between how cash-strapped small biotech companies operate and how big pharma throws money around.

Our interactions were solely with people from Lederle Laboratories, a wholly owned subsidiary of ACY that was responsible for drug development. Lederle Labs were located in Pearl River in the Hudson Valley in New York State. It had a massive campus that employed over three thousand people at that time. Rural New York State is dotted with these industrial campuses, surrounded by their colossal parking lots and neighboring high-end hotels to house their guests.

Our team meetings sometimes took place at Lederle. On those occasions we were picked up by limos from LaGuardia and driven the thirty-plus kilometers to the Pearl River Hilton. We'd be picked up in the morning by limos and transported to Lederle and the sumptuous boardrooms where we met. Meals were provided in the private dining room that served senior management.

I had never been so busy in my life. I had a full teaching load and a thriving lab at the university. I had been invited to sit on the immunology grants panel for the Medical Research Council and on another grants panel for the National Cancer Institute of Canada. The research on BPD was expanding. If BPD was going to be the next PDT product, we had to understand the molecule thoroughly. Anna Richter and I felt we had to understand the kinetics of the drug in the body, even down to sub-cellular locations. We also had to figure out how to formulate the drug in preparation for its use in humans. In addition to the BPD research, I also had a number of students in my lab doing projects in basic immunology and cancer immunology.

At the same time I had to get familiar with a whole new field – PDT with Photofrin. I started going to specialized PDT meetings. And I had to get on friendly terms with Tom Dougherty, the founder of Photomedica and developer of Photofrin. I knew that Tom regarded us with suspicion, and I didn't blame him. In his shoes, I would have been suspicious too. Anna and I took a trip to Buffalo to meet with him and soon we were on good terms. Tom is a soft-spoken man of gentle temperament, who over the years has shown a fierce loyalty to the people around him. Underneath his gentle exterior, he is a dogged idealist with strong belief in his work. He had carved out a powerful position for himself at Roswell Park, a research facility associated with a cancer hospital. Tom maintained a remarkably strong relationship with the physicians he worked with. A PhD chemist, he had worked out how to collaborate with physicians and succeeded in building his own specialized PDT surgical suite for treating patients.

Experimental approaches to cancer treatment are not uncommon in institutes like Roswell Park. During the 1970s, Tom developed his concept that light-activated drugs could have a place in cancer treatment. Initially, Tom thought that Photofrin did not need to be submitted to the FDA for permission to conduct early human trials in New York State. It was manufactured in New York and was being used there, so it did not cross state lines. The FDA thought otherwise. With the cooperation of some of the physicians, he obtained an investigational new drug permit from the FDA, a document that legitimized his using Photofrin to treat patients.

PDT is a complex procedure. It involves administration of the drug, followed by light treatment (in the case of Photofrin, forty-eight hours after drug administration). The light used with Photofrin is at 632 nm. This wavelength is red light, like the LEDs we use in many signals such as traffic brake lights. If the light for PDT is to be administered internal to the body, say to a tumor, it has to be delivered via fiber optic cables, which resemble clear plastic tubes. One can see the light traveling through the cable like fluid. In that case it is best for the light to be very tightly focused onto the cable. This means that laser light is the best, and in some cases, the only suitable source. That's because lasers can emit light that is spatially coherent, which allows for maximal concentration of the beam and for minimal power loss. PDT is a multi-disciplinary activity involving biologists, physicists, and physicians who are familiar with lasers and cancer treatment.

The only lasers that delivered light at 632 nm at that time were massive research tools two to three meters in length. They tended to be unreliable, so the surgical team had to include a laser physicist.

The eventual wide acceptance of PDT would require the development and distribution of reliable, reasonably sized and priced lasers.

Tom was able to bring together people who were courageous enough to pioneer these kinds of experiments. Many of the physicians Tom worked with were doing fellowships at Roswell Park and moved on to take up clinical positions in other parts of the United States or in other countries. Many of them became convinced of the validity of PDT and continued to use Photofrin wherever they went. By the mid-eighties, there were PDT users dotted throughout the United States, Japan, China, and Europe, the pioneers who championed the treatment.

When Johnson and Johnson bought Photomedica two years earlier, they moved PDT beyond an experimental therapy. By that time Tom's apostles were worldwide, and Tom was making his "bathtub Photofrin" and shipping it around the world. With the advent of Johnson and Johnson's involvement, company-sponsored phase II clinical trials were undertaken until the company decided to jettison the project, at roughly the time I gave my seminar in Hamilton.

Tom made a point of seeing that I got invitations to attend clinical meetings at which PDT clinicians got together and shared their results. I was really surprised to see that specialists from most medical disciplines were represented: dermatologists, pulmonologists, head and neck specialists (otolaryngologists), urologists, gastroenterologists, gynecologists, and ophthalmologists, all convening because of their belief in the technology. All these early investigators shared a pioneering spirit, and I enjoyed their presentations and their company. Most of them were real characters with a touch of cowboy about them. One of the leaders among them was a pulmonary surgeon named James McCann. James was a big rangy man and a chain smoker. He always had a cigarette on the go when he gave talks and I heard that he smoked in the operating room when he was delicately delivering a fiber optic to a patient's lung cancer.

On another occasion, I heard an ophthalmologist who made presentations on treating ocular tumors with Photofrin. At the end of his talk, he put in a plug for PDT to be adopted more widely for treating ocular conditions. He drew the analogy between cancer and other ocular conditions like age-related macular degeneration (AMD),

Me, my mother, and sister
Pamela on our way to Canada,
August 1940

Graduation from UBC, 1955

My favorite place: in my lab doing tissue culture

The cabin on Sonora surrounded by cow parsnip in bloom

Skunk works: the first of my experiments with PDT using plant sap from
cow parsnip and sunlight on a wart

QLT founders (l to r): John Brown, Tony Philips, me, Jim Miller, Bruce Hay
(our accountant), and Ron McKenzie

David Dolphin and I with a molecular model of our drug, BPD-MA

FDA approval of Photofrin for esophageal cancer: me with
Dr. Charles Lightdale, one of our key investigators and
spokesperson

FDA approval of Photofrin: me with Tom Dougherty (the father of PDT) and
Karole Sutherland

John North, our chief scientific officer, at the IPA annual meeting in Nantes

Receiving an award from the Foundation Fighting Blindness (l to r): Ian Harper, me, Ed, Kate Butchovsky, and Bob Butchovsky

Receiving an honorary doctorate from the University of Ottawa, 1993

Cake honoring my retirement from UBC as professor emerita

With my mother on her 90th birthday

With Ed and Lucy

my mother's condition. He pointed out that the selectivity by which photosensitizer molecules concentrated in tumors was probably attributable to the new blood vessels that form around growing tumors. These vessels, termed neovasculature, are frequently faulty and tend to leak fluid. He pointed out further that many ocular pathologies such as AMD are also attributable to neovasculature. That got my attention, and I filed the information away in my brain for a later date.

Towards the end of 1988, I was invited to give a talk at a conference organized by the Wellman Laboratories, a wing of the Harvard dermatology department. I was invited to specifically talk about our work with photosensitizer-antibody conjugates. It was a small conference and I felt a bit out of my depth, because many of the talks were on complex photochemistry and physics, unlike the meetings I'd attended with clinicians. But my talk went well.

After I had spoken, I met the woman who had organized the meeting. Her name was Tayyaba Hasan. Tayyaba was a faculty member at the Wellman and at Harvard. Her background was in photophysics, but she had become interested in the conjugation work and was starting to do some research in the area. I was delighted to find a kindred spirit, and we became good friends. We continue to be friends. Our interests over the years have coincided on many occasions, not least of which was to collaborate in the early ocular work that led to the first treatment for age-related macular degeneration.

CHAPTER TWELVE

The Nation's Blood Supply

QLT underwent huge changes between 1989 and 1991. I was awarded a university-industry fellowship from the Medical Research Council that paid half my salary. The other half came from QLT, where I was spending more and more time. I remembered my vow to myself when I became a faculty member that I would never lecture from "yellowed lecture notes," and I realized I was in danger of becoming one of those professors. I simply didn't have the time to prepare properly for my lectures. With this award, my UBC salary was liberated, so the department could hire another faculty member to take on my teaching duties. The only responsibility I had at the university was for my research lab and my students. I could handle that.

As we moved forward with BPD, preparing it for clinical trials, we hired more scientific staff at the company, people who understood scale-up and drug manufacturing. I learned the complexity of preparing a new chemical entity for the clinic. In research, chemists like David Dolphin can create new chemical structures at the lab bench. The amounts created under lab conditions are always small. David would send us a few milligrams of a new chemical. We would carefully weigh it and dissolve it in dimethyl sulfoxide (DMSO), a chemical known to be safe in animals and very good at dissolving hydrophobic compounds (substances that are not soluble in water). However, it is not approved for use in humans, so it could not form part of the drug product. DMSO is used in the freezing of cells and tissues because it protects the delicate walls of cells from being pierced by ice crystals during the freezing process. It is used to store cells to be used in bone marrow transplants, and therefore is used in humans for these limited purposes, but it has never been put through rigorous testing, nor received regulatory approval for broader use.

For clinical development, kilogram quantities of the experimental drug are required. Scaling a procedure from grams to kilograms can be very problematic and is never seamless. The manufacturing process has to be honed to be as simple, reproducible, and as foolproof as possible. Every additional step required in the synthesis of the product involves some loss of the active ingredient, so simplicity is highly desirable.

The chemical steps to creating BPD seemed relatively straightforward. The starting material for both BPD and Photofrin was heme, the small molecule that binds to the hemoglobin molecule and forms the molecular "pocket" that carries iron around in the blood. Heme does not vary in struvcture between different animal species, and so can be sourced from any blood supply. Blood from cows or pigs is a plentiful and cheap starting material. However, mad cow disease had recently become a problem, particularly in Europe, so we had to make sure our starting material did not come from cow's blood.

Heme is put through a number of chemical steps to convert it to BPD. BPD is a very hydrophobic molecule, so it is not readily dissolved in physiological solutions. Therefore it was necessary to formulate BPD into liposomes, which complicated the development path. Liposomes constitute a formulation of the active ingredient with fatty molecules that assemble as tiny nanoparticles that can easily be administered as an injectable. ACY scientists helped us create a liposomal formulation that worked extremely well.

The chain of BPD production ended up being complicated. A company in Holland produced and sold us the heme. A company in Edmonton did the chemical modifications to produce the active BPD molecule. It was shipped to Japan, where it was formulated into the liposomal product. The bulk liposomal material was then shipped to a company in the United States where it was filter sterilized, bottled, and freeze-dried. Then the bottles had to be QC/QA checked and labeled. Samples were taken at every step, checked for purity, and stored for reference purposes and for stability. All these steps have to be taken under Good Manufacturing Practices (GMP), which necessitates the extremely detailed and extensive code of regulations that govern production of drugs. Doing things under GMP adds to the complexity and expense, but is, for the most part, reasonable.

We went through this arduous process because we were partnered with ACY and were learning to do things the way they are done in the pharmaceutical industry. It was costly (in the millions), but in retrospect I believe we did the right thing.

Many small companies have neither the expertise nor the money to follow these kinds of procedures when they're trying to get a new drug into the clinic to see how it performs in humans. After all, the drug might fail, and then all the money needed to get the drug into the right formulation would be wasted. The FDA does permit, under some circumstances, clinical testing of a new drug not produced under GMP, so long as the agency is sure the product is safe. This is an argument I've heard many companies make to avoid these costly steps at the early stages of drug development. "We just have to get it into the clinic to test it," would be the argument. "We can go back and do the appropriate manufacturing steps later, when we have data." And this is true. Many start-up biotechnology companies have trouble raising sufficient funds to carry out proper drug development steps. Getting a product into clinical testing can well be a financial inflection point for a start-up, enabling them to get investor interest and money. Unfortunately, any biotech company that does its preclinical toxicology and early clinical work on a product that is later reformulated may be told by the FDA that they have to go back and repeat all the toxicology work with the final formulated material. While this may be the only way possible for a start-up company, big pharma would regard this approach costly in terms of time lost.

In addition to our work with ACY, in 1989 QLT signed a collaboration agreement with Baxter Healthcare. This large pharmaceutical company focuses almost exclusively on blood and blood products. Its largest customers are blood banks and hospitals. Baxter supplies them with the bulk of the disposable plastic materials used in the process of blood collection, testing, and treatment. In the late eighties the

HIV scare was escalating, and the first hemophiliacs who received blood products for their condition were testing positive for AIDS and Hepatitis C. Soon after that, cases showed up in patients who had received units of blood during surgery. Until this happened, AIDS was considered a disease exclusively found in the homosexual community and was thought to be transmitted sexually. The first cases of AIDS occurred in North America in the early eighties, but the blood-banking crisis did not occur until the late eighties. This was because those first victims might not have become symptomatic for several years after infection, during which time they might have been blood donors, unaware of the deadly virus in their blood. Blood banks were in turmoil and scrambling to get adequate testing of blood from donors. Baxter was heavily involved in this search.

We had scientific reasons for thinking BPD would be a realistic candidate for viral inactivation in blood products. The wavelength at which BPD is activated is a wavelength near the infrared part of the spectrum at 692 nm. Light at this wavelength does not generate a lot of heat, but it penetrates animal tissue very well. This is significant when considering the various ways in which the drug can be used. A simple demonstration of the power of red light in the range of 690 nm to penetrate tissue is to hold a full spectrum flashlight to your finger, palm, or ear lobe in a darkened room. The only light that you will see passing through the tissue is red. The white light is called full spectrum because it contains all the visible wavelengths of light. When it is shone into tissue, the only part of the spectrum that gets through is the red part, the rest being absorbed and refracted by tissue.

This property of BPD meant that it could possibly be used to eliminate unwanted cells or viruses in blood. We became interested in this application of PDT because it had potential in cancer treatment. Autologous bone marrow transplantation was being quite widely used at this time as an approach to treating some leukemias and lymphomas. In these diseases, patients are treated initially with chemotherapy. When the patient goes into remission, their bone marrow can be drawn and stored frozen. In cases of relapse, they could receive very high dose chemo- or radiation therapy at levels sufficient to kill the patient in the absence of rescue with healthy bone marrow. The problem with this approach, in addition to its being extremely risky and hard on the patient, is that it frequently fails because of residual cancer cells in the bone marrow that was withdrawn from the patient. When this marrow is reintroduced, the patient is being given cancer. These failures led to research into ways by which bone marrow taken from a patient could be purged to rid it of residual cancer cells while sparing the crucial stem cells.

We looked at BPD as a possible purging agent and found that we could selectively kill significant numbers of chronic granulocytic leukemia (CGL) cells in bone marrow preparations while sparing the essential stem cells. In addition, we found that viruses in blood were highly susceptible to being killed with BPD. Their very small size meant that it took only low doses of the drug to kill viruses in comparison to human cells. Because of the ability of 692 nm light to penetrate materials containing hemoglobin, we realized we might be able to treat red blood cells and rid them of viruses including HIV and the Hepatitis C viruses.

Baxter was heavily involved in research into purging agents. People from Baxter heard one of my students present our data on purging leukemia cells at a scientific meeting and made contact. We negotiated a collaboration agreement. The agreement covered a number of salaries at QLT as well as Baxter's access to the drug BPD for use in blood and blood products. We undertook to look at the feasibility of using BPD to render safe an assortment of blood products, including red cells and platelet preparations. There was not a lot of money in the deal with Baxter, but it gave QLT prestige to have two large pharmaceutical partners. Our shareholders liked the deal and rewarded us by increasing our stock price.

The people we collaborated with were mainly engineers and scientists in one of Baxter's research facilities at their enormous campus in Deerfield, Illinois. We got on very well with our Baxter counterparts. They were scientists, like us, interested in problem solving. We all enjoyed the work. It was new and interesting. The team worked fast and well together. Baxter was primarily interested in making platelet preparations safer, so we started out working with both platelets and red blood cell preparations. Platelets are the blood product that is heavily used by hemophiliacs. These patients are unable to clot blood. A relatively minor injury can be life-threatening. A minor blow can cause internal bleeding. The quickest way to stop bleeding is to inject platelets. Hemophiliacs were the canaries in the mine, harbingers of the panic to come over the nation's blood supply with AIDS and Hepatitis C.

Baxter had great interest in developing a process to render platelets safe. Platelets are separated from whole blood by differential centrifugation. The bulk of cells in blood are the hemoglobin-containing red cells, and they can be separated easily from the plasma by centrifugation. The white cells constitute a minor population in comparison to the reds and co-precipitate with the red cells during centrifugation, forming a discrete layer on top of the denser red cells. Platelets are left in the liquid plasma after red cells are removed. They are much smaller than the red and white cells and far more fragile, but are critical for blood clotting.

We found that we could kill seven to eight logs (ten million to a hundred million) of virus particles per ml in red cell concentrates with no damage to the red cells themselves. The expected shelf life of BPD-treated red cell concentrates was not significantly different from that of untreated samples. We were elated.

Results were not so encouraging with platelets. While we were able to kill viruses selectively in platelet preparations, the magnitude of the effect was much less. Platelets are small and fragile and easily damaged. We struggled with the platelet situation, trying to find a better way of delivering drug and light.

But Baxter was keen to go ahead with blood treatment. Our engineering colleagues at Baxter started figuring out how PDT could be inserted into the process of blood collection and treatment. We brought in the American Red Cross to help us model how this new procedure could be incorporated into the process. Blood collection involves blood being drawn as single units from the donors, tested in various ways, and processed to separate the red cells from plasma and to prepare platelet concentrates or other blood products. If BPD were to be added to units of whole blood, it would have to be added prior to separation of the cells from the plasma. Then the unit of blood would have to be treated with light. This would entail transferring the blood to another container through a narrow aperture and passing it under a light source. A model of how this process could be incorporated into blood collection was constructed. It turned out that this extra step would add something like $40 to a unit of blood (about a third of the actual cost). Everyone started to get worried about the viability of the project.

Baxter swallowed hard, but didn't go negative. The next step was to go to the FDA and float the idea of what they referred to jokingly as "getting approval to contaminate the nation's blood supply with an unapproved drug."

The FDA's reaction was neutral to negative and there was no enthusiasm for the project. Our claim was that we could remove millions of virus particles from blood, even though such levels of contamination could never possibly exist in any blood sample. We had used feline leukemia virus as a virus similar to HIV, because we couldn't get HIV-infected blood samples with virus counts as high as we wanted to test. The FDA was unimpressed. Their counter regarding HIV was that even if we were to treat HIV-infected blood so that there was no longer any detectable virus, that would not be conclusive evidence that there was really no virus present. Absence of proof is not proof of absence. HIV may be harbored inside certain white blood cells in a dormant form. The virus in such a form would not be detectable by methods available at the time. So while we could claim to make blood safer, we could not guarantee that it was completely safe. They went further to say that there would soon be far better detection methods for both HIV and Hepatitis C virus available, so the need for a blood "disinfectant" would not be so pressing.

We all knew that this opinion was the kiss of death. Our friends at Baxter were sad but pragmatic. These things happened. Senior management at Baxter terminated the partnership. Big companies make decisions on killing projects all the time, and there are no repercussions for the company. Employees are assigned new projects. But when this happens to a small company, the effects can be devastating.

In the first place, we had to put out a press release saying that our partnership with Baxter was over. Our shareholders were not happy and the stock dropped. Also, I had to tell the scientific staff that the project they had poured their energy into for the past eighteen months was being axed. I thought long and hard about how to tell them. I realized I couldn't just go to them and say, "The project just got killed and I'm not sure what you'll be doing now." I figured I should already have a plan as to which projects they'd be assigned to.

I worked hard on trying to put people into other projects that would not make them feel displaced and that would suit their talents.

People in science become very devoted to their projects, and can be grief-stricken when their work is terminated. People feel belittled and undervalued in situations like this and I was very sympathetic. As I had anticipated, there was a lot of pain and anxiety about their positions in the company, but we got through it.

Well, I thought, at least we have the ACY partnership. We'd better make sure that stays on track.

I remembered talking to a friend of mine soon after we signed the agreement with ACY in 1987. I was going on about how exciting the deal was and how nice the people at ACY were. His warning to me was, "There's one thing you can be sure of when you're dealing with a big company: No matter how good your relationship with them is now, you can be very sure it will be very different in five years and not nearly so good." In due course, this was proven to be true.

CHAPTER THIRTEEN

PDT Is Different from Other Cancer Drugs

During 1990, we at QLT all learned a lot about phase III clinical trials from our Lederle partners. Stu Marcus was the Lederle clinical leader of the Photofrin project team. He was an MD-PhD with a specialty in oncology and held himself in high regard. He was also a strong believer in PDT, so despite his initial condescension towards us he was a terrific ally. We didn't take his condescension personally. It was his standard way of behaving.

Before we acquired Photofrin from Johnson and Johnson, the drug had been used extensively in humans, first by Tom Dougherty, who had, after a rough start, obtained permission from the Food and Drug Administration to do so. He had obtained an investigational new drug application (IND) from the FDA that allowed him to test the drug in cancer patients and to carry out what are called phase I clinical trials. To get an IND, an investigator or a company has to provide enough information to the FDA to show that the drug, used under the conditions proposed, will most likely be safe in humans. This proof is provided by submitting what is called a preclinical package, which summarizes all the toxicology work performed in appropriate animal species as well as information on manufacturing and all the preclinical research carried out using the drug.

Phase I clinical trials in cancer drugs are usually designed to show the drug is safe to administer to patients, by starting with a very low dose of the new drug and gradually escalating the drug dose until there are any signs of toxicity, at which time the drug will have reached its maximum tolerated dose (MTD). Cancer drugs are usually tested in patients who have failed other approved therapies and are considered terminal. Often there is little expectation of a therapeutic effect, especially at the starting dosage.

Once a drug has been shown to be safe, it moves to phase II clinical trials. These studies involve larger groups of patients in defined groups with a specific disease at a given stage. These patients are selected from those who have had recurrences following primary therapy. They are sorted into cohorts and given different dose regimens of the test drug to better define optimum effects. Phase II trials are often comparative with another treatment regimen in order to compare the efficacy of the new drug with a known standard of care. However, they are usually "open-labeled," which means that all involved know whether the drug being administered is the test drug.

PDT is different from other cancer drugs. The photosensitizer itself is not toxic in the absence of light, so estimating an MTD is not necessary. Light activation at the tumor site will cause local toxicity, so trials start by administering low light doses and escalating thereafter to achieve the greatest anti-tumor effect while sparing surrounding normal human tissue.

Johnson and Johnson had already performed phase II clinical trials with Photofrin in a variety of types of cancer patients. When QLT and ACY became involved, over five hundred patients worldwide had received Photofrin therapy. Many of the doctors who had worked with Tom at Roswell Park had continued to use PDT after they finished their residency with him. These physicians operated under what are known as investigator-sponsored trials. No strict protocols were followed, so the data acquired were unreliable. On top of that, Johnson and Johnson had conducted a number of phase II trials at several sites. But although the data were a bit of a dog's breakfast, ACY's Lederle Laboratories decided there was sufficient evidence of benefit to proceed to phase III trials.

Phase III trials are also called registration trials because they are the final studies that lead to applying to the FDA for approval to market a new drug. They are the most costly part of drug development trials and always involve comparative studies, either against the standard of care for the condition being treated or against a placebo if there is no approved standard of care.

At Lederle's insistence, we undertook to carry out phase III trials in three types of cancer: bladder, esophagus, and lung. The way in which big pharma designs pivotal trials is to draft up a rough protocol as to how the trial will be run, then call together clinicians from across the continent who are willing to participate in the studies. These clinicians meet to thrash out the fine details of the protocol. In recruiting willing physicians it is important to include doctors known to be key opinion leaders (KOLs) in their field, who will be essential in helping to promote the product once it comes to market.

One of the KOLs is chosen to be the principal investigator. In the case of Photofrin, the KOLs included mainly Tom Dougherty's converts who were out in the world using PDT to treat patients already, so they had access to lasers and experience in the technology. But they were seasoned physicians, and these meetings, led by Stu Marcus, were sometimes raucous and acrimonious affairs, with the KOLs overriding Stu's positions. Stu managed this room full of egos as large as his own with surprising skill.

After settling on the trial design, we went to the US FDA to seek feedback on the acceptability of the designs. Pharmaceutical companies unofficially consider the FDA the toughest regulatory body in the world. The agency controls distribution and sale of drugs to the most lucrative market in the world, so it is not surprising that FDA buy-in is regarded as essential for drug development. Health Canada and regulatory bodies in other jurisdictions would be approached once the FDA had approved the protocols.

There is a standard way of approaching the FDA, something as ritualistic as an ancient Greek facing an oracle on Mount Olympus. The "sponsor," the company that owns the drug and is in charge of the clinical trial, has to frame its queries as carefully worded simple questions that can be answered by a simple yes or no. They usually start with "Does the agency consider ..." or "Would the agency agree ..." A great deal of thought and preparation goes into these FDA meetings, as well as rehearsal, since the sponsor is permitted to give a brief presentation prior to asking the questions. Failure to satisfy the FDA, while not fatal, can cause months of delay.

To obtain FDA approval to market a new drug, it is necessary to run two well-designed and controlled clinical trials for a given disease indication, where what counts as an indication is usually a relatively narrow subtype of what we ordinarily call a disease. So in the case of cancer, an example indication would not be bladder cancer, but something like "stage III recurrent bladder cancer *in situ*." Most companies would not consider conducting phase III trials without first getting FDA buy-in that the protocol was suitable and that their end points will be acceptable for approval of their drug. The two trials can be identical or they can be different, but they must be run at different locations and managed separately. The ideal trial design involves the classic double-blind study, in which the test drug is compared either to the approved standard of care or, if there is no such approved drug, to a placebo. In a double-blind study, a patient receives either the test drug or a placebo under circumstances in which neither the patient nor the treating physician is aware of which patients have received the new drug. With Photofrin PDT, it was impossible to carry out a blinded study because of the lasting skin photosensitivity, so we had to find a way around this dilemma in a manner that would satisfy the FDA.

We proposed treating bladder cancer patients with recurrent carcinoma *in situ*, a condition in which superficial cancer cells spread throughout the inner lining of the bladder. For PDT, balloon catheters attached to a fiber optic were inserted into the bladder via the urethra, inflated, and treated with light shone on the whole bladder wall. The second pivotal trial in bladder cancer was to treat patients who had had recurrent bladder cancers after partial bladder removal and were at the stage where they were candidates for cystectomy (removal of the bladder). These patients were referred to as having "beaten-up bladders" that couldn't take another surgical intervention. The end point for PDT would be to delay or prevent cystectomy.

In lung cancer, the decision was to run one large trial in the United States, Canada, and Europe, plus a second smaller one. For the larger study we planned to treat terminal lung cancer patients with recurrent local progressive disease. These patients were often treated with YAG lasers (a laser with a crystal made of yttrium, aluminum and garnet) to clear obstructive disease. YAGs are thermal lasers and simply burn through the tissue. They can cause perforations of the lung and fatal hemoptysis (bleeding from the lung). YAG lasers have been known to occasionally cause fires in the operating room. We hoped to show that PDT offered a safer and more tolerable treatment than YAG. The second study was to be a small one, offering PDT to patients diagnosed with early, superficial cancer in the lung. Normally, patients presenting with early lung cancer are candidates for surgery as a potentially curative procedure. However, some patients with this condition are in such poor health with lung diseases like chronic obstructive pulmonary disease (COPD) or emphysema that they are not candidates for surgery. PDT could be offered to this population as an alternative. The patient population with these characteristics is very small, considering that only 5 per cent of lung cancer patients are diagnosed with early disease and only a small percentage of them would not be candidates for surgery. However, for this tiny population, PDT offered a potential cure for their disease.

In esophageal cancer patients, we would compare PDT to YAG treatment to clear the esophagus. I remember that during a project team meeting, after the protocols had been put together, Ed asked whether it would be possible to get FDA approval to run a non-comparative clinical trial on patients for whom there were no treatment options – patients who had completely blocked esophagi and, for technical reasons, were no longer candidates for YAG therapy. Stu Marcus rolled his eyes and said in a very condescending tone, "I find that question touchingly naïve." He went on to say that if Ed wanted to make that suggestion to the FDA, he should do so, but no one from Lederle would back him up. The irony of this moment was that Ed did put the question to the FDA when we went there with the protocols. The meeting itself was well attended, because this was a new cancer treatment and people in the oncology division of the agency were interested. Bob Temple, the head of the division, was there. When the question was asked, Bob Temple responded: "Give me data on twelve patients with completely blocked esophagi who can drink a beer after your treatment and I'll approve the study as one pivotal trial," he said.

This did go forward as one of the esophagus studies and was eventually approved. Stu conveniently forgot his original comments. Having the second pivotal trial as non-comparative probably shortened the process by many months if not years.

Small biotech companies never have the resources to run such extensive clinical trials as we did with Photofrin. We were following the pattern set by big pharma and we were stuck with this costly responsibility. They call this approach having "multiple shots on goal." The money we had raised was far short of what we would need to pay for this extensive program. ACY agreed to start sharing the costs of the trials. Nevertheless, Jim, our CFO Rashid Aziz, and I had to go back to the market to raise more money. Our bankers, Nesbitt-Burns, now part of the Bank of Montreal, were willing to take us on the road and brought in some US bankers. Our story was compelling; we had three phase III trials under way and a deal with Baxter. We hit all the major banking locations: Toronto, Montreal, New York, Boston, San Francisco, London, Paris, and several others. Biotechnology was in its salad days and we were able to raise forty million dollars.

Once the clinical trials with Photofrin got under way towards the end of 1990, the project team settled down to get the work done. There were additional toxicology studies to be carried out to satisfy the final requirements the FDA would have for Photofrin. The paperwork associated with clinical trials is formidable. The data from each patient entered into a study will cover dozens of pages. Each data point entered has to be double-checked for accuracy and signed off.

In addition to all the clinical and preclinical work, there is another huge and costly part of drug development, namely the chemistry and manufacturing. Historically, poor manufacturing was one of the key motivators underlying FDA regulation. There had been disasters where citizens had died as a result of poorly manufactured or adulterated medications. As a result, the regulations concerning manufacturing were excruciatingly detailed and unforgiving. On balance, I thought this was a good thing.

Lasers continued to be an issue. Ed, who had decided after his year of sabbatical at QLT that he would resign from the university and become an employee at QLT, was now in charge of regulatory affairs and lasers in addition to being the lead of QLT's project team. It was he who pointed out to us that ACY/Lederle was not paying enough attention to the laser issues. It seems they thought of PDT as an addition to oncology cancer chemotherapy. The reality was that PDT was quite distinct. It was not a technique that would be taken up by oncologists; it was a procedure that would be used by surgeons and endoscopists familiar with interventional treatments. An operating theater with access to specific lasers would be required. And the lasers being used by our KOLs were research lasers, not the sort of apparatus that would be usable by a broader base of clinicians.

Ed became a voice in the wilderness, trying to get the attention of Lederle, pointing out the need to act soon on upgrading lasers or making a deal with a laser company to start producing PDT-appropriate lasers that didn't require constant tweaking. We hired a couple of laser specialists at QLT.

While we were making strides with Photofrin, we were also moving forward in the lab with BPD. We completed the preclinical development with no hitches. The drug was now properly formulated and had passed all the toxicology requirements. We started planning the first phase I clinical trials with it and I was thrilled at the prospect. We decided we would treat skin cancers, either skin cancer itself or metastatic disease that had spread to the skin. The post-mastectomy patients with local recurrences seen in Hamilton were an example of the second. We fixed on an intravenous dose of BPD that we knew was very safe. We calculated the time interval needed for the drug to reach high enough levels in the skin to have an effect when activated with light, based on careful animal studies.

PDT is a complex procedure: not only does one have to calculate a dose of the drug, one has to establish an optimal light dose as well as an optimal time after administration of the product to achieve the maximum effect. Because we knew that the effect of light versus drug concentration with BPD was reciprocal (the more light you use, the proportionally less drug you need to achieve equivalent effects), we could do a study keeping drug dose constant while titrating the amount of light used to treat lesions. At Stu Marcus's insistence, we agreed to administer the BPD as a slow infusion from a blood bag and treat three hours after the start of injection.

The BPD phase I trial was to be run at the Wellman Labs, where my friend Tayyaba Hasan was a faculty member. The Wellman is part of Harvard's department of dermatology and is dedicated to photomedicine. The faculty included physicists, chemists, laser specialists, biologists and dermatologists. It was an ideal location for a phase I clinical trial. Rox Anderson, a PDT specialist and dermatologist, was selected as the chief clinical investigator for our phase I trial.

It was time for BPD to be christened. ACY submitted an application to the USAN (United States Adopted Name), an organization that remains mysterious to me to this day. Its machinations involve cooperation with other groups to come up with generic names that are descriptive and simple for new chemical entities. BPD became verteporfin. I could see the connection to green porphyrin.

I was ecstatic about how well BPD had performed as we prepared it for clinical trials. Anna and many students in my lab had worked so hard on this drug. We had studied its behavior under many different conditions and were confident in its potential. It is hard to put into words the thrill that we all felt about seeing BPD take this next giant step. Something we had all played a part in creating was actually going to be tested in humans.

CHAPTER FOURTEEN

I Was Surprised and Not at All Pleased

Although the project team had meshed successfully by the end of 1991 and we were moving forward with both Photofrin and BPD, friction was growing between QLT and ACY/Lederle management. Management of both companies met together every quarter to assess the progress of the team. Jim Miller, our CEO, had become impatient with the way things were running. Progress was a bit too slow for him and he didn't like how much everything cost. He liked action and deal making. He also liked being at the wheel. The senior people at Lederle Laboratories were all pretty much out of a mold – used to being bosses, well-mannered and civil, never in a hurry or tempted by short cuts. They were accustomed to things being done according to standards they set. They were conservative. I could tell that Jim irritated them by the way he pushed boundaries and challenged their specific rights to Photofrin when he got the opportunity.

A key person we hired at this time was Karole Sutherland, a woman of exceptional organizational skills who had a chameleon-like ability to fit into many different roles. Originally trained as an OR nurse, she had over the years assumed responsible administrative roles in a number of projects, including the world's fair in Vancouver, Expo '86. She was very competent and was soon filling the clinical role in the Photofrin team. She got on well with both Stu Marcus, the Photofrin clinical leader, and Lou Gura, the Lederle project manager.

Jim had hired a couple of people to take on the position of VP clinical at QLT, but they had proven to be duds. Karole, whose sardonic smile always made me wonder what she was plotting, endured these people. Fortunately, she has a wicked sense of humor.

Bad hires are a big drain on a small company. They cause a loss of time and create anxiety. They can also be embarrassing. So we were all delighted when Jim found a very personable candidate for VP clinical. His name was Barry Steiger, and we all thought highly of him. He was knowledgeable and low-key. He was from the United States and came to Vancouver ahead of his wife, who had a career of her own. Jim had assured them both that there would be no problem with her pursuing her career in Canada, and that he would take care of the paper work to get her a work permit.

Jim and Rashid Aziz, our CFO, were also having their differences. I knew there was conflict but never inquired into what the problem was. This was the spring of 1991 and the time for annual raises, but I wasn't paying much attention, as I was far too busy with the work on BPD in my lab and at the company.

My salary was divided between my MRC industrial fellowship and QLT. When our letters from Jim came around with our salary for the next year, again I paid no attention. Money has never been a big motivator for me. But when Rashid came to me one afternoon and told me that Jim had purposely lowered my QLT salary instead of giving me a raise as promised, I was incensed. I asked Rashid if it had been a mistake. He shook his head.

I knew Jim pretty well by this time. I was angry; I didn't like being taken for a sucker. I knew with Jim that this wasn't sexism, it was one-upmanship. My pay statements were complicated. My MRC statements were funneled through the university and had a number of deductions on them and my statement from QLT also withheld money for tax and RRSPs, so what I got and the total pay were very different. Jim knew me and probably figured I wouldn't do the math. And he was right.

Our board of directors had expanded recently. We had added Jan Dlouhy from ACY, Duff Scott and Peter Cosgrove from the financial sector in Toronto, and Bob Feeney, a retired executive from Pfizer Pharmaceuticals in New York. In addition, there was Ray McLean, who had helped us financially earlier, and we five founders. With

these new board members, we also started having board committees, a main one of which was the compensation committee.

As with most companies, Jim as the CEO decided raises for senior
management at QLT and then the proposed salaries were submitted
to the compensation committee. The compensation committee had
approved my salary, so I couldn't just go and confront Jim about the
injustice he'd done.

I thanked Rashid and wrote an angry and sarcastic letter to the chair
of our board of directors' compensation committee, Peter Cosgrove.

The letter written by Jim announcing our salaries stated that the
change in salary would be retroactive to the beginning of the year,
about five months earlier. In my letter to Peter I asked if that meant
I'd have to pay back the difference between the two salaries. After
I'd sent my letter to Peter, I went to Jim and told him what I had
done and gave him a copy of my letter.

I could see that Jim was uncomfortable. He apologized laughingly
for the "error" he had made and said we could have straightened
things out without going to Peter. I pointed out that the salary had
gone through the board already, so Peter had to be involved.

Peter was furious with Jim when he got my letter. He phoned
me, apologized, and said he'd had no idea I had been treated in this
way. Jim had said everyone had salary increases. Peter said he was
coming to Vancouver to look into a few other things regarding the
leadership of the company.

Peter Cosgrove arrived in Vancouver and stayed a couple of days.
He spent time with Barry, who was becoming very upset when he
realized that getting his wife a work permit was more problematic
than had been presented.

A special board meeting was called by Duff, to take place in
Toronto. I was very nervous about the meeting. I knew it was going
to be ugly. Just before I left, Barry came to me and told me he had
resigned; he was fed up. I tried to dissuade him, but he was ada-
mant. I wondered how the board would react to this news.

Soon after, Jim tendered his resignation and I was asked to become
interim CEO of the company. I had assumed this role would go to
Rashid, so I was surprised and not at all pleased. I didn't mind going
to Wall Street and talking about the science, but CEO? I knew I was
regarded by the investment community as an egghead and was

happy with the designation. And I was a woman. Worse, another potential investor had referred to me as "that hippy woman." So I knew I wouldn't be good for the company as its CEO. Women CEOs in the biotech industry were non-existent at that time. But I accepted the role, under the condition that we would pull out all the stops to find an appropriate candidate for CEO. I would make that my highest priority.

The collaboration with Baxter had just come unstuck and now QLT had me as CEO. It was a daunting thought.

We hired a headhunter and I crossed my fingers. Duff Scott, our board chair, and I spent a lot of time interviewing possible CEOs. I don't remember how many, but enough to make me depressed. One interview in particular stands out. It took place in New York and we were using Bob Feeney's office. He was a retired executive at Pfizer and still had an office there. The candidate was a man in his forties, and he horrified me; the term "snake oil salesman" kept popping into my head. The man made me squirm.

At the end of the interview, Bob took the candidate out to the reception area and was gone a few minutes. I turned to Duff and said something to the effect that this man seemed so sleazy I thought we'd hit rock bottom. Duff was non-committal, so I really didn't know what he'd thought. Bob came back smiling and said, "Wasn't that fantastic. What an interview."

I gaped at him. "You're kidding," I said. I really thought he was being droll but realized he wasn't. I wondered if we'd been in the same meeting.

Towards the end of the year, I was starting to despair about finding a CEO, but Ed came up with an idea. He had visited the head of regulatory affairs at Cyanamid Canada about the Photofrin trials. While there he had met Bud Foran, the CEO of the Canadian subsidiary. Bud was about to retire and planned on moving to Vancouver from Toronto. Ed had a flash of insight: why not see if Bud would be interested in steering QLT for a couple of years, helping us through this rocky period and taking his time finding the right successor? I loved the idea, and our motley crew of senior executives was game for this possibility. Bud was a genial, seasoned manager and he was aligned with our pharmaceutical partner. Bud liked the idea, as did our board and the ACY management. We quickly arrived at a suitable arrangement. I heaved a big sigh of relief and went back to the science.

Bud Foran was an exceedingly nice man. He was the picture of a CEO – tall, white-haired, well-spoken, with impeccable taste in dress, so good that we wondered if he'd been a model at some point in his career. But he'd come from a marketing background and had only a vague idea about drug development. Cyanamid Canada, like most pharmaceutical subsidiaries, was concerned mainly with product sales, and these included fertilizers and an assortment of chemicals along with pharmaceuticals. If clinical trials are done in subsidiaries, they are usually late-stage confirmatory studies that do not require a big staff of clinical investigators. Even then, the trials are always directed from head office.

Despite Bud's lack of experience in the details of what we were doing, he was a definite asset to us. The investment community approved of him. He looked the part and affirmed the solidity of our relationship with ACY.

And Bud brought more than just a suitable public image. He understood the appropriate infrastructure for a small company and made it his priority to address the management needs we had. Within the company, there was a general feeling of relief. A new calm prevailed and people got on with their work.

So 1991 ended on a positive note, despite the reverses we'd experienced that year. I looked forward to 1992, the year BPD would be tested in humans.

CHAPTER FIFTEEN

Splitting Up

During my years as a researcher I have been thrilled when a crucial experiment has worked out exactly as I had hoped, confirming a theory I'd had. But nothing I had experienced came close to our first patient treated with BPD. She was a breast cancer patient with a recurrent tumor on her chest wall, a particularly difficult skin lesion to treat and one that added insult to the injury of a mastectomy. I wasn't present in Boston to witness it, but Rox Anderson, the dermatologist running the trial, was in touch with us, taking pictures. The patient suffered no ill effects from the drug and the light treatment administered three hours later was tolerated without any pain. We of course were starting with a light dose we considered sub-optimal based on our animal data. The patient was followed carefully for a number of days and, as predicted, her levels of circulating BPD subsided fast and she had no skin photosensitivity measurable after two days. At about forty-eight hours the lesion became inflamed and darkened visibly. Within a couple of weeks it had become scabby and all but disappeared. We were overjoyed. Even at the low dose, there had been an effect. All the careful animal work was paying off, and our calculations had been very predictive.

It was wonderful to work with the Wellman staff. These people were top scientists, working together in the multi-disciplinary activity that constituted photobiology. Rox was a good scientist as well as a physician, and that made a big difference in developing optimal treatment parameters. We expected the phase I work to continue for at least a year, amending the protocol as we learned more about optimizing the treatment.

But QLT was running low on money. The clinical trials we were supporting were expensive. And Bud Foran wanted to create a solid infrastructure at QLT. We went on the road again in the spring of 1992. This was my first road trip with someone other than Jim as CEO. Bud had no experience talking to investors and had to be coached. His presentations were always scripted for him, and he deferred to Rashid or me on any questions that involved our development programs. It was far easier than the ad hoc follow-ups I had to do following Jim's presentations or responses. Bud was courteous and somewhat self-deprecating. Road shows are physically and mentally exhausting. He had never been subjected to anything like a road show before and he was sixty-seven, so he got tired. I was ten years his junior and I know how worn out I got. We were often up for 7:00 a.m. meetings and then kept going until early evening, sometimes covering a lot of ground, often in different cities. When Jim, Rashid, and I had done these junkets, we always got together at the end of the day for a glass of wine and a good meal and to let our hair down. We usually laughed a lot. Our dinners with Bud were more formal and not nearly so relaxing.

I remember the final day when we closed the financing in Toronto. We had raised thirty-four million dollars, so we had sufficient funds to last us a couple of years. I felt good about where the company was. We had stable management and Bud had done well on the road.

I took a flight to Milan that evening to attend an International Photodynamic Association meeting. This was an international body of PDT clinicians and scientists. They came together to present data on clinical outcomes, hear reports on experimental new photosensitizers, and discuss progress. Our Lederle colleagues were attending and we planned a team meeting during the congress.

I presented a report on BPD and on our first few treated patients. There was some excitement amongst PDT practitioners about our new photosensitizer, BPD. I was leaving the lecture hall when an attractive young woman stopped me and introduced herself as Ursula Schmidt. She was an ophthalmologist from Lubeck, Germany and was currently doing a fellowship at the Wellman with my friend and collaborator, Tayyaba Hasan. I hadn't met Ursula and was surprised that Tayyaba had an interest in ophthalmology.

We went for coffee and she told me she was working on an approach to treat age-related macular degeneration (AMD) with PDT. I got excited. She then said she wanted to do some proof-of-concept work in rabbits using BPD. I got really excited. We had been supplying Tayyaba with unformulated BPD for her research, but Ursula said she wanted to use the liposomal formulation that we were using in our phase I cancer trials.

She confessed that she had been hanging around the clinic when Rox was treating his cancer patients with liposomally formulated BPD and taking the drug remaining in the infusion bags when the treatment was finished. The drug was being used at 2.0 mg per kg of body weight, so depending on the patient's weight there would be different amounts of drug remaining in the bags. Such an act is frowned upon by the industry. Unused drug product is supposed to be accounted for before it is disposed of. The Wellman, being a research facility rather than a hospital, turned a blind eye if another investigator wanted to use the dregs. When I asked Tayyaba at a later time why she had never thought to tell me she was involved in research on eye conditions, she said she'd had no idea I would be interested.

I was impressed by Ursula's initiative and thrilled she wanted to work on AMD. I told her I had a long-time interest in the potential for PDT to treat neovascular diseases in the eye because of my mother's condition. Neovasculature, or the formation of new blood vessels that do not mature properly and leak, is the basic cause of AMD. My mother was now legally blind from the disease. I told Ursula I would be happy to supply her with formulated BPD. I asked her what proof-of-concept work she was doing. She said she was using a tumor model in the eyes of rabbits, realizing that the growing tumors would cause neovasculature to develop in the same region of the eye as AMD did. She knew it wasn't an ideal model but said she had hopes of being able to progress to a well-recognized primate model after finishing the rabbit work. Her goal was to work out the treatment parameters (dose, timing, light dose) in rabbit eyes for closing the neovasculature and then use that information to carry out work in the primate model. Her plan made perfect sense. Primates are very expensive, so the more preparatory work that can be done in other species, the better.

I liked Ursula immediately. She had a dry sense of humor and an intensity about achieving her goals as soon as possible that impressed me. I had long hoped that QLT would at some time get a chance to work on AMD, and now this opportunity had presented itself. The timing was perfect. We had just raised money, so we could certainly support some research at the Wellman on testing the feasibility of ocular PDT for AMD. I'd have no trouble justifying spending the money to support this research.

By the fall of 1992, Ursula had completed the rabbit work and we were ready to think about the primate work. Ed and I flew to Boston to meet with Tayyaba about how to set up the next stages. Ursula's fellowship had finished, and she was back in Germany but hoped to return to Boston soon. Tayyaba had recruited another ophthalmologist named Joan Miller, who was currently a junior faculty member at the Massachusetts Eye and Ear Infirmary (MEEI), Harvard's ophthalmology department. Joan had access to the monkey model, the accepted animal model for AMD, and was interested in working with us. We set up a collaboration with Tayyaba, Ursula, and Joan that covered the dose ranging and treatment of induced neovasculature in the eyes of macaque monkeys.

I must admit to having concerns about primate work. I suppose I'm an animal chauvinist to some extent. While I never minded working on mice, I had misgivings about anything larger – even rats, which seemed so intelligent. And I had always held firm that under my supervision no animals suffered pain or discomfort; but I felt squeamish about primates. The reason we had to carry out the work on primates was because the primate eye closely resembles the human eye, whereas the eyes of other animals do not.

The model involved creating areas of neovasculature on the back (retina) of the eye using a quick pinpoint thermal laser treatment. Once the neovasculature formed, the animals would be treated with PDT to see if that treatment successfully closed off the neovasculature. Animals would be anaesthetized during the procedure. I observed a treatment and was satisfied that the animals were being treated with humaneness, still had vision after our procedure, and were not suffering pain.

It became clear even at early stages of this work that our idea for treating ocular conditions with PDT had promise. The study was expanded in order to prove safety as well as efficacy to the FDA, so we had to continue building a strong database, slowly testing timing of treatment and light dosing. While I was tremendously interested in this project, it was definitely not mainstream at QLT. We were a cancer company, working with a big pharmaceutical partner that had a focus in oncology.

Bud, as our new CEO, continued to fulfill what he saw as his mandate to create a solid infrastructure for QLT. We were now a phase III company and Bud saw the necessity of hiring a marketing person. He hired a young woman who had been doing market research for a large cosmetics company in New York, but had worked in the pharmaceutical industry at Pfizer. Her name was Lee Anne Pilson, and she had met a young Canadian lawyer during a Club Med vacation. They had married and she had moved to Vancouver.

I had an exceedingly naïve idea about marketing. I couldn't see a role for Lee Anne at QLT. After all, we weren't selling anything. I found her an exotic species with her elegant style of dress and perfect manicure and makeup. I made an effort to get to know more about what marketers did. In retrospect I know Bud was absolutely right in hiring her. We had a drug in phase III, and while ACY would be responsible for sales of Photofrin it was important for QLT to have someone in the company who could relate to ACY marketers and possibly have significant input into marketing strategy, including how the relationship between clinical development and future marketing would develop. Over the years I came to admire Lee Anne's keen analytical intelligence and to appreciate the importance of marketing intelligence.

Bud also created proper investor relations (IR) and human resources (HR) departments. He persuaded Ed that he should move to corporate development rather than stay in regulatory affairs. Ed hesitated, because he enjoyed the FDA interactions, but agreed in the end, largely because he recognized the competence of Alexandra Mancini, an experienced regulatory affairs person he had hired. In retrospect Ed realized that Bud had been right about suggesting this change. He became the VP of business development.

Bud's final hire in management was a chief operating officer, intended to be Bud's successor. His name was Randal Chase, and he had been a junior executive at a Canadian subsidiary of a major pharmaceutical company. He had a PhD in biochemistry from UBC and had earned his stripes in a big pharmaceutical company in drug development. His credentials seemed perfect. I admired Bud for having provided our company with a solid base of the kind of expertise we needed to get through the next few years.

Alex Mancini was made VP of regulatory affairs, taking over Ed's former position, and we hired Andrew Strong as director of clinical affairs. Andrew was a PhD in pharmacology and had worked in clinical affairs at Bayer. He and I interacted a lot and respected each other. Andrew was not on the Photofrin team and worked mainly with the BPD researchers. ACY had rights to BPD in the field of cancer but not in other disease indications such as AMD, so we spent time looking at other applications for PDT and running animal models to investigate possibilities. We had the ocular research going on at the Wellman, but at QLT and my university lab we were investigating the possibility that PDT could be used in diseases like psoriasis and rheumatoid arthritis. Andrew acted as the intermediate between research and clinical.

The company seemed to be on an even keel until towards the end of 1993. Then trouble loomed. ACY announced that it was going to purchase Immunex, a big Seattle-based biotech company that had a number of cloned and engineered products in development for the treatment of cancer, the most promising of which was Pixikine. They also had an approved product, a cytokine called Leukine. Immunex was to merge with Lederle Laboratories to form an enlarged research facility for ACY on the west coast. Immunex would retain its location and name. This decision, in my opinion, had a significant impact on ACY's subsequent decision about Photofrin.

Mergers are always disruptive. Departments are forced to review their programs and weed out what may be seen as redundancies. The oncology division at Lederle was excited about the merger. There were promising new drugs coming under their purview in the future. The drug Pixikine was in phase III trials for cancer and looked like it had the potential to become a blockbuster. It was an engineered dual-purpose molecule and a departure from traditional cancer drugs. The only other cancer product Lederle had in development was Photofrin, and there were early signs that Photofrin did not have blockbuster potential. Lasers were problematic.

Lederle's sales force that called on oncologists would also have to start calling on endoscopists to market Photofrin.

The truth was that ACY senior management was distracted by the merger with Immunex and PDT was becoming a nuisance. The first thing that happened in this regard was a serious discussion about the bladder cancer indication. In 1990, the FDA had approved another treatment for superficial bladder cancer. This treatment was the intravesicular infusion of the bacillus calmette guerin (BCG) vaccine.

BCG was developed in the 1920s as a vaccine against tuberculosis (TB). BCG is a suspension of attenuated TB bacteria (mutated so that they have lost their ability to cause disease). It was an inexpensive generic vaccine in the 1990s. Introduction of BCG into the bladder causes a significant inflammatory response. Individual investigators had been using BCG to treat bladder cancer in this way since the 1970s, and there was a lot of information in the scientific literature about BCG's ability to eliminate superficial bladder cancer cells, presumably by immunological means during the inflammatory response to the BCG. By 1990, a cooperative group of oncologists had run a clinical trial and produced sufficient information on the efficacy of BCG to eliminate these cancer cells to warrant an FDA approval. BCG was the first approved immunotherapy for cancer. No major pharma company was marketing BCG when it was approved. BCG vaccines were generic and available at a low price from companies like Connaught Labs. In 1990, ACY didn't consider BCG a threat to Photofrin's projected market share, but by 1992 the treatment was gaining traction. Improved products were being produced and more doctors were adopting the therapy.

As an immunologist, I had always been very interested in following the BCG results in bladder cancer. It was the first real evidence produced in clinical studies that indicated that stimulation of the immune response locally could be instrumental in combatting cancers. As I write this book twenty-five years later in 2018, immunotherapy for cancer using check-point inhibitors is providing real hope that cancers can be defeated using the body's own immune systems.

ACY used this situation to conclude that they were no longer interested in pursuing the bladder cancer indication with Photofrin. By this time we had completed the bladder trials and had submitted an application for approval in Canada. ACY didn't care. They were not going to market Photofrin for bladder cancer. Superficial bladder cancer was too small an indication and not worth ACY's trouble.

Then results started coming in from the lung and esophageal trials. Bob Desjardins, Stu Marcus's boss, threw up his hands at the list of adverse events reported for both Photofrin and YAG treatments in late-stage lung and esophageal cancer. He said ACY would never be able to market Photofrin with a package insert listing all those adverse events. For all pharmaceutical products marketed after FDA approval it is necessary to list all adverse event thought to be associated with the therapy in those package inserts, which come with the product in miniscule print.

The patients in the phase III trials were very sick people with terminal disease. One would expect adverse events. With Photofrin there would always be prolonged photosensitivity. One patient reported a severe sunburn on her face after treatment when she went to a mall and stood admiring a large Christmas tree illuminated with red lights. It was unfortunate that the red lights of the tree activated the Photofrin residing in her skin. On the other hand, I was glad she was well enough to go to the mall. It was clear to us that ACY wanted to dump the program.

We waited for the other shoe to drop, and it did, early in 1993. At the next management meeting between ACY and QLT, the ACY people said they wanted to dissolve the partnership. We would get back our rights to Photofrin and BPD. It was clear to me that the ACY management felt bad. We said we intended to file for approval for both lung and esophageal cancers with or without ACY. All the data from the clinical trials had been documented and stored at Lederle. ACY management very kindly offered to help us transfer all data to our own databases, and even offered assistance from Lederle regulatory and statistics staff to put the FDA filing together. They weren't obliged to do that. Alex Mancini and Karole Sutherland continued to work with the Lederle staff through 1994 and into 1995 as they prepared the documents for filing for marketing approval with the FDA.

The only part of the partnership that remained intact was with Lederle Japan. The Japanese subsidiary had a degree of independence from the head office, and the Japanese continued to have an interest in PDT. This continued interest would come back to haunt us in the future, but we were grateful at the time that part of the relationship continued.

I didn't feel anger towards ACY. I was disappointed. They had made decisions based on what they saw as sound business models. And they were bending over backwards to help us through this difficult time. Our investors would punish us. But for ACY, Photofrin was a nuisance and a distraction from their Immunex blockbuster. Ironically, Pixikine failed to reach its endpoints in its phase III studies. Its trademarked name was abandoned in 1993 and it was forgotten.

Soon after that, in 1994 ACY was taken over by American Home Products and also disappeared. And we were on our own.

CHAPTER SIXTEEN

Failure Was Never an Option

I suppose some people in our circumstances would have given up. But whether it was brainless stupidity or sheer bravado on our part, failure was never an option for us. Photofrin PDT was a treatment option that could benefit many people, and we had the responsibility of bringing it through to approval. We felt we were so close to getting Photofrin approved in the United States. We had completed the phase III clinical trials, and even filed for approval in Canada, Japan, and Europe. Indeed, we were much closer than we had been a couple of years earlier, but there was still an enormous amount of work to do.

Our ACY colleagues had put a lot of clinical data together in preparation for filing with regulatory bodies. They had performed a lot of statistical analyses on that information. ACY management gave Photofrin back to us after their regulatory people had filed for approval in Canada and Japan. Margaret Dugan, one of ACY's senior regulatory people, was enormously helpful to Alex Mancini and Karole Sutherland at this time. She and Lou Gura orchestrated the transfer of the mass of data accumulated at ACY and helped in preparing the documents for filing with the FDA. .

We had bonded with the ACY team. Those ACY staff closest to the project, Stu Marcus and Lou Gura, dealt with the project termination differently. Lou was a seasoned large pharma employee. He was a bit of a cynic and certainly did not let himself get emotional about any project he might lead. He was typical in that way of most people I have met in large pharma companies. The biggest difference between those in big companies and those in start-ups is passion or lack thereof for projects they are involved with. We at QLT cared desperately about the work we did. We believed in it and wanted it to succeed, not for monetary reward but for the sake of doing something good. I'm sure those kinds of feelings do exist in big pharma, but they are not as readily apparent. The people at ACY involved in the hands-on development work were used to senior management making snap decisions about canceling projects, so they didn't get as invested emotionally.

Stu Marcus was the exception. He was relatively new to big pharma and he did care passionately about PDT and, as a clinician, believed in it. I think he was heartbroken when ACY dropped the project. As he moved forward in his career, he took positions in small companies focusing on PDT, always faithful to his first love.

Also, we were painfully aware that we looked like a failure to the investment community. We had lost two significant partners, ACY and Baxter, and appeared to be clinging to a failed technology. But we didn't think we were. There was a place for PDT. We just had to keep our heads down and make progress on both the Photofrin and BPD projects. But we had to achieve some pretty obvious but tough milestones before we could win back our credibility on Wall Street.

We were lucky to be in a fairly good financial position. We had more than thirty million dollars in the bank. We figured we needed about three years to get our credibility back, and that money would have to last until then. We would certainly have to go back to the market before we became profitable. The cost of preparing a filing for Photofrin in the United States for was not huge in comparison to what had already been spent on the development. The expensive part of the work was complete, since the clinical trials were over. The next steps would take time but not a lot of money. Early-stage research, like what we were doing with BPD, was also not excessively expensive. However, once we made a decision as to the development path for BPD, we would need more money to cover the costs of clinical trials and manufacturing.

But we had to achieve the milestones we set ourselves. First, we had to successfully gain approvals for Photofrin in major markets. Not only did we have to gain approvals, we had to find ourselves suitable marketing partners or make the decision to market the drug ourselves. Bud and Lee Anne would have loved to see us market Photofrin ourselves, since both of them came from a marketing background, but caution prevailed. It would be far too expensive and distracting to hire a marketing force. We had to make the money we had last until we could go back to the stock market.

Our primary goal, we decided, was to seek a marketing partner in the United States and Europe while we worked on our filing in the US. Ideally, we would find one international company that would market the drug worldwide. Our vice president of marketing, Lee Anne Pilson, went into high gear. She worked with us all to put together a compelling slide deck to sell Photofrin marketing rights to a suitable partner. We targeted companies that had PDT research programs and a cancer franchise.

We couldn't afford to take BPD further in clinical trials, so we continued to support the ocular program in Boston and carry out exploratory research on different animal models for a variety of diseases.

In late 1993 we got our first Photofrin approval in Canada for the treatment of superficial bladder cancer. It was a meaningless approval. We couldn't market the drug ourselves and the BCG vaccine had been well accepted for this indication. Both treatments (BCG and Photofrin) caused a lot of discomfort for patients, because they caused inflammatory responses in the bladder. Photofrin's local negative effect was of shorter duration than the BCG effect, since BCG had to be administered repeatedly over a month or so and patients experienced discomfort for the duration, but the skin photosensitivity caused by Photofrin was a disadvantage and lasted six to eight weeks. Also, BCG was cheap and didn't require a laser. In early 1994, to our surprise, Japan approved Photofrin for the treatment for early lung cancer in patients who were not candidates for surgery. Lederle Japan was responsible for marketing Photofrin there, but decided not to actually launch the product until they got an approval for late-stage cancer. A product launch is an expensive affair, and companies won't launch a product until they have a good chance of big profits from sales. The number of early-stage lung cancer patients for whom this treatment is suitable is very small. Lung cancer is rarely diagnosed early (between 5 and 10 per cent of all cases) and, of those, at least half will be candidates for surgery. And while PDT is a valuable option for those patients who are not eligible for surgery, large pharma companies don't get excited about launching a product for such a small population. The first European country to approve Photofrin was Holland, in April, 1994, and there we were approved for both lung and esophageal cancer.

We took to the road, visiting a number of pharmaceutical companies in Europe, large and small, looking for the right fit. Partnering with large pharmaceutical companies is an arduous and time-consuming affair. The big companies are never in a hurry to make a decision, whereas we felt pressure to get a deal done. A deal would give us back some of our lost credibility and probably enhance our dwindling finances.

PDT was a new technology, and companies that had a cancer franchise were familiar and comfortable with oncology drugs that they marketed to oncologists. But PDT needed laser expertise and lasers. We received a lot of polite refusals.

Then in the fall of 1993, we contacted Ciba-Geigy in Basel, Switzerland, and they agreed that we could come and present to them. Ciba was one of the large pharmas that actually had a PDT research program and a photosensitizer already being used by investigators in patients. The drug was a phthalocyanine, a photosensitizer that, in our opinion, did not have many of the favorable characteristics of BPD. We were very hopeful when we got to Basel, a beautiful old city close to Zurich.

We pitched the market potential of Photofrin and covered the early research we were doing with BPD. Much to our surprise, when we started to discuss future possibilities it turned out that Ciba wanted to close down their PDT program and hoped to license their photosensitizer to us. What we didn't know at the time was that Ciba was preparing to merge with Sandoz, another Swiss pharmaceutical company, to form Novartis. We were all too familiar with jettisoning of inconvenient projects from our experience with ACY. It seemed like PDT was an inconvenient technology for big pharma. And what we didn't need was another early stage sensitizer that wasn't as good as BPD. We had had high hopes for a partnership with Ciba, so we were very disappointed.

When the meeting had started we had asked the Ciba business people whether they represented all of Ciba – that is, all therapeutic areas. At first they said yes, but then added that they did not speak for Ciba Vision, the ocular subsidiary of Ciba-Geigy. As we were preparing to leave, we said we would be interested in talking to Ciba Vision, and asked whether that could be arranged, since we had some interesting early-stage work in ophthalmology. They said they would set that up for us when we were next in Europe,

Back in Vancouver, I met with Lee Anne Pilson, our marketer. We talked about the possibilities for PDT to treat age-related macular degeneration. She set to work on the research as to whether there was a sufficient market to justify a development program. Lee Anne's work on AMD was my first introduction into the value of market research. I was impressed.

Lee Anne was also impressed when she ran the numbers on wet AMD. She found out that there were 500,000 new cases of wet AMD per year in the United States alone. And there was no satisfactory treatment for this debilitating disease. Her research went well beyond just pulling these numbers off the internet. I got a crash course in the analyses performed to get net present values. It certainly looked like AMD was an ideal indication to pursue for BPD.

In 1994 we made contact with the management at Ciba-Geigy and asked if they could arrange a meeting with Ciba Vision. They convened a group in Basel. Our delegation included Ursula Schmidt, who reported on her work at the Wellman in Boston. The key person from Ciba Vision was Gusti Huber, then Ciba Vision's VP of international R & D. Gusti was a person we all took to immediately. He was open, intelligent, and easy to get along with. And from the first presentation he was enthusiastic about the potential for PDT to be used in macular degeneration.

The follow-up meeting was some months later at Ciba Vision's headquarters in Bulach, about fifty kilometers from Zurich. Bulach was not much more than a farming village and the Ciba Vision offices were not impressive. They were located on the upper floor of the Swiss version of a strip mall, above a gumboot shop. The company itself specialized in contact lenses and eye drops. Their main operations were located in Atlanta, Georgia, so the Bulach location was established to accommodate only the management and European marketing and sales group.

Gusti was familiar with the PDT project within Ciba-Geigy and was extremely interested in Ursula's rabbit work and the studies we had been supporting in Boston. We met the other management people at Ciba Vision. The CEO, a man called Luzi von Bidder, was affable and showed enthusiasm for PDT.

We started talking. During 1994, we took many trips to Bulach. Both Ursula Schmidt and Joan Miller accompanied us on separate occasions. During that year, Ciba Vision decided that they wanted to initiate a PDT project. But they were getting internal pressure from their colleagues in Basel to adopt the Ciba phthalocyanine drug rather than BPD, even though no preparatory work had been done in the eye with their drug. I think that both Luzi and Tony Ellery, Ciba Vision's business development person, leaned towards their own drug because they would not have to pay for it.

But Gusti Huber was a scientist, and he scrutinized the data we had generated and the solid scientific basis we had for making the claims we did about our drug. Gusti became a staunch ally of BPD and QLT within Ciba Vision for the right reasons. He became so committed to the program that he asked to be reassigned from VP of research to being the manager for the project. Gusti was one of the few people I knew in big pharma whose careers were firmly linked to their company but who showed the kind of commitment he did for a project. I also learned how critical it is for small companies partnering with large companies to have an internal champion. We had had one in Stu Marcus at ACY, and now we had Gusti.

The discussions about how to structure a deal went on for the better part of a year. Luzi negotiated throughout to try to get the best deal he could for his company. He was very good at playing at being hard-up.

Back at QLT, Rashid left the company. I don't know what had happened, but I was sorry to see him go. He'd been a solid friend to me. Rashid had hired a young recently minted chartered accountant named Ken Galbraith into our finance department. When Rashid left, Ken became the chief financial officer. Ken had an extremely incisive mind and became a key strategist for QLT.

I didn't know what to make of our COO, Randal Chase. I wasn't able to develop any feeling of rapport with him. I wondered how I would feel about reporting to him once he became CEO when Bud retired. I tried to figure out my feelings but was unable to. And that troubled me.

We had gone through a lot of growing pains in getting a management team that functioned. Under Jim, there had been a succession of inappropriate hires that didn't last long. Bud had been very constructive in understanding the kind of management we needed and had cobbled together a group that worked reasonably well together. But we were a pretty mixed bag. We all had our foibles. Our management team was fairly gender balanced, and that made management meetings interesting. Our females, Lee Anne, Alex, and myself, were no shrinking violets. Then we had Bud, Ed, Ken Galbraith, David Main (head of investor relations), and Randal. While I might not choose to spend a lot of time with some of these people, I could relate to them all on a human level and appreciate what they contributed – except for Randal. I tried and failed.

With input from the Wellman and MEEI in Boston, and Andrew Strong, Anna, and myself at QLT, we put together the information needed to get FDA approval to do a phase I/II trial in patients with macular degeneration. These would be patients with what is called the wet form of macular degeneration, which means they had neovasculature (new blood vessels) in their maculae, the part of the eye critically responsible for central vision.

The progression of AMD is thought to start with the dry form of the disease, which may convert to the wet form as the disease worsens. Dry AMD is characterized by residues of fatty materials called drusen at the back of the eye. It is extremely common in people over the age of eighty and doesn't usually cause significant vision loss. It is thought that drusen consist of accumulated waste substances that should clear of their own accord but do not do so in the aging eye. New blood vessels can form around drusen in an attempt on the part of the body to clear the deposits. The new blood vessels do not mature properly and can leak into the back of the eye, resulting in the condition of wet AMD. Eventually scar tissue forms, which results in severe and irreversible vision loss.

Our goal was to close the leaky vessels with PDT. We knew the drug accumulated preferentially in the walls of neovasculature. Activation by light could effectively destroy those vessels, while leaving normal vessels unharmed. The primate work had shown we had a big safety window under conditions that successfully closed new blood vessels at the back of the eye. By the end of 1994, we had put together an information package and a trial proposal to send to the FDA. We would use the same approach we had taken for the first clinical testing of BPD for people with cancerous lesions on their skin.

The original protocol for BPD had involved a forty-five minute infusion. Ophthalmologists tend to work fast and don't want patients occupying space in their offices for long periods of time, so faster infusion time was necessary if we were to have a successful product. In this study, the drug would be administered within ten minutes and light treatment would take place fifteen minutes after the start of infusion. We would start with a very low light dose.

We finally signed an agreement with Ciba Vision. Luzi had gone on about how his R & D budget was cut to the bone, how he couldn't squeeze another penny out of head office for an up-front payment. He sang the blues with great effect. We let him go on and then let him have his way: there would be no up-front payment for rights to BPD in ophthalmology. But we got him to agree that Ciba Vision would pay 60 per cent of the development costs while we paid 40 per cent. Marketing rights would be Ciba's exclusively worldwide and we would share profits fifty-fifty after expenses. I'm sure Luzi thought he had made a good deal.

We had agreed when we started negotiating that we wanted a co-development arrangement with both companies paying for 50 per cent of development costs. Our demand for an up-front payment was to level the playing field. We had already spent a lot of money on BPD and expected to get at least fifty per cent of that back in an up-front payment from Ciba.

It was Ken Galbraith, Ed, and our corporate lawyer, Hector Mackay-Dunn, who put the agreement together. I don't think Luzi had ever been involved in doing clinical trials for developing a new chemical entity. Most ophthalmology drugs were first used in other therapeutic areas and adapted for ocular use. In addition, Ciba Vision made contact lenses and eye drops, and development costs for products such as these are trivial in comparison to the costs of taking a new drug through to worldwide approvals.

By the time we had completed our phase III trials, his extra 20 per cent of the development costs far exceeded the kind of up-front payments we would have considered fair and that Luzi had fought so hard to avoid. We would probably have accepted an up-front of between ten and twenty million. Then we would have shared further development fifty-fifty. Luzi thought he had made himself a great deal, but we knew more about what our clinical trials would cost than he did. We were very happy with the agreement.

CHAPTER SEVENTEEN

Nothing Can Prepare You for the Torture of ODAC

After signing the agreement with Ciba Vision, we formed a joint development committee and a joint management committee. Gusti, the former vice president of R & D at Ciba, became project director.

Our first priority was to get BPD into phase I/II trials for macular degeneration. Since we had already done phase I studies for cancer, this trial would encompass both safety and dose ranging. We could have done a straightforward phase I to establish the safety of treating eyes, but the FDA often accepts a combination study in which both safety and efficacy are established.

Andrew Strong wrote the protocol with a lot of input from Joan Miller, Tayyaba, Anna Richter, and myself.

We initiated the trial in 1995. It was run at the MEEI by Joan Miller and at the ophthalmology department in Lubeck, Germany, where Ursula Schmidt had returned after her fellowship in Boston. They were both experienced with the technique and had suitable lasers. Both groups were in touch with the Wellman.

It wasn't long before Joan had her first patient, willing to be patient zero. He tolerated the treatment with no difficulty. However, he did vomit a short while afterwards, and that concerned us briefly. Joan said he had become very nervous when he heard he was patient zero for this trial and she was sure the cause of his nausea was nerves. He'd also apparently consumed a large hamburger and chips immediately before the treatment. Joan's diagnosis seems to have been right, because we had no other case of nausea following treatment.

The patient returned for follow-up fluorescein angiography the next day. This procedure involves intravenous injection of the fluorescent dye fluorescein. Fluorescein remains sufficiently long in blood vessels to provide an opportunity to track blood vessels at the back of the eye using a special camera that photographs the distribution of the fluorescein. This procedure is used to diagnose wet AMD because the fluorescein leaks out of the damaged blood vessels and can be seen as a pool of fluorescence on the retina. Similarly, closure of those vessels will also be shown by an absence of the fluorescent pool and indeed an absence of any fluorescence in areas where blood vessels have been closed.

The treated eye of the first patient showed clearly that his patch of leaky vessels was closed by the treatment. There was no fluorescence in the area that had been treated. We were overjoyed.

What remained to be done in our phase I/II trial was to repeat the treatment on a cohort of patients, and then follow them to see if the vessels stayed closed and to determine if there were any late-occurring ill effects of the treatment. The next cohort would be treated with an escalating light dose to determine tolerability and endurance of the treatment effect in order to find out the optimal regimen. Ursula Schmidt started treating patients in Lubeck very soon after patient zero.

We worked well with the Ciba project team. Gusti was a pleasure to work with and he always made the interest of the project paramount. Also, Ciba Vision was a kind of orphan group within the large Ciba-Geigy organization; its corporate culture was more like that of a small company. But they had little experience in drug development. We, on the other hand, had expertise in PDT and in running drug development trials. It was a good feeling that now we were the ones bringing our skills to the table as opposed to our position with ACY, where we were the novices.

However, things were about to change in Vancouver. Bud called me into his office and told me that it was time for him to retire. He said Duff had advised him to do so. He said that, after all, he had hired Randal as his successor. Randal had been at QLT almost a year now. I said I was sorry. Could he change his mind? Everything was going so well. He said no, he wanted to work on his golf game. And so now Randal was our CEO.

Operationally, nothing much changed. We were all very busy. Then we had an interaction with Proctor and Gamble. They were interested in PDT. We were surprised. I always thought of Proctor and Gamble as producers of soaps and detergents. But apparently they were seriously considering building up their pharmaceutical business and wanted to look at PDT as a technology platform. I think it was Randal who made the initial contact, and he was very excited about the possibilities. Most of us were excited about the possibilities of this interaction, but cautious. Partnerships with large companies were complicated. We had seen that with American Cyanamid. Then Proctor and Gamble made it clear that if they decided to go forward with PDT they would likely buy the company and want all rights to BPD back from Ciba. That was a blow. We had forged a strong working relationship with Ciba Vision and the ocular opportunity was very promising. An acquisition of our company would undoubtedly disrupt that relationship.

We had meetings with the Proctor and Gamble business development people. They were nice enough folk to deal with but it was apparent that their lack of experience would present serious difficulties to a technology as science-based as PDT. Also, we had our agreement with Ciba Vision. We were doing a clinical trial with them. Proctor and Gamble emphasized that a condition of a deal with them would be breaking off our relationship with Ciba. Randal was very keen on closing the deal, but I was convinced that a Proctor and Gamble acquisition of QLT would destroy the company. The research would be halted and QLT would lose its science focus.

Then Randal decided to go to Switzerland to have a meeting with Luzi von Bidder, Ciba's CEO. Ken and Ed were worried that he was going to try to foul up our agreement with Ciba Vision because of the Proctor and Gamble discussions. I wasn't sure how much I trusted Luzi. If Luzi could extract payment for giving up on the BPD project, I believed he'd do it in a heartbeat. We caucused without Randal and decided that Ed would accompany Randal.

Ed made his arrangement without Randal's knowledge and just showed up at the airport saying he had to meet with Gusti. Ed's presence helped preserve our relationship with Ciba.

I concluded that Randal was more enthusiastic about the Proctor and Gamble deal than the rest of us in management. He was used to working in a large company and had been successful in that environment. Also, he had not been at QLT while we forged our relationship with Ciba. If QLT were absorbed by Proctor and Gamble, such a change wouldn't upset him. Ed, Ken, David Main, and Lee Anne all felt that such a transaction would put all the PDT programs in jeopardy. Certainly, the ocular program would be terminated. We all felt strongly that we should make every effort to prevent such an outcome. We decided to keep very close tabs on Randal's actions, to ensure they would not prejudice our ongoing programs.

In addition to all this drama at the company, we continued to make a concerted effort to complete the Photofrin filing with the FDA. The responsibility lay with Alex Mancini, our vice president of regulatory affairs. She is a perfectionist and she wasn't happy with the data collection, especially from the European sites. She worried that there were too many gaps in patient follow-up. Her counterparts at ACY encouraged her to submit what was there, so, with misgivings, she went ahead and filed the data she had.

Filings with the FDA require that the sponsoring company provide detailed information on every patient treated, their vital statistics, and their follow-up information. It is not unusual for data on a single patient to fill a four-inch-thick binder, especially patients as sick as the ones we were treating. Then there is the copious number of statistical analyses of the data for each individual endpoint. It is possible now to submit these filings electronically. In 1995 that wasn't possible. It was not unusual for filings to contain up to a thousand large binders.

We submitted the Photofrin filings for lung, early- and late-stage disease, and for esophageal cancers, both the single-arm study and the comparative study, in early 1995. The filing is called a new drug application (NDA). There is a filing fee the company has to pay in order to offset the costs of the review process. At that time it was in the range of a hundred thousand dollars, but it now exceeds a million. After you file, you are at the mercy of the FDA. Usually, in the oncology division, the review waiting time is in the range of six to eight months.

There are various responses you can get from the FDA. The least desirable is one is called a refusal to file (RTF). The FDA, if it is going to issue an RTF, has to do so within sixty days of filing. An RTF means that the filing is seriously flawed and the drug is not approvable as the data package stands. We didn't expect an RTF.

The best you can hope for is a date being set for a meeting with an advisory committee, in this case called an Oncology Division Advisory Committee or ODAC. This means that the reviewers at FDA have considered that the file is on the path to approval and the final major hurdle is to convince the committee that approval is appropriate. Not all drug approvals involve an advisory committee, but any new chemical entity or new kind of therapy will almost invariably involve one.

FDA can also send you an "approvable" letter. This is not good news. This means that your file is deficient in some ways that are considered remediable. In the letter the FDA outlines the deficiencies in the submission. It may mean that the sponsor has to go back and do more clinical work, or provide additional information. An approvable letter most likely means that the approval will be held up for months if not years.

We got a letter with an advisory committee date. Alex had done an impressive job, putting together a credible filing with the data set she had. She went into high gear in preparation for the company's appearance.

ODACs are a regulatory person's trial by fire. They are not held behind closed doors. They are open to the public and can turn into a circus. They are heavily attended by investment bankers who are looking for promising investments and analysts who are looking to pick winners or to pillory losers. They are also attended by the press and various advocacy groups who have an opportunity to speak if they wish to do so.

The company is invited to make a short (no more than thirty minutes) presentation, painting the most positive picture of the drug, followed by questioning from the FDA reviewer or members of the advisory committee. Then the FDA reviewer gives a presentation that frequently is critical in pointing out weaknesses of the trials and other aspects of the filing. Then the advisory committee members get to question the FDA reviewer or to provide their own criticisms or comments on the data. During this part of the proceeding, the sponsor is not allowed to say anything. So no matter how off course you might think the discussion is getting, you have to suffer in silence. Then you get to the moment of truth, when a series of questions are put to the committee. One or more of the questions ask whether or not the committee recommends that the drug be approved. A show of hands from the committee can seal the fate of a small company like ours.

The ordeal is not over even then, because the FDA has to decide whether it accepts the advisory committee's recommendation. The FDA considers all the information provided by the company and presented at the ODAC and after due deliberation will inform the company where they stand. The FDA doesn't necessarily follow recommendations from an ODAC, so the company is usually kept guessing for a couple of weeks more before they hear from the FDA. The whole process is agonizing and can be publicly humiliating.

Investment analysts attended our Photofrin ODAC from the Canadian banks that had financed the company. Tom Dougherty came with others from Roswell Park. There was an advocacy group there, patients who had benefitted from the treatment as well as many people I didn't recognize. Alex had rehearsed us relentlessly. We had a mock FDA reviewer, played by Lou Gura, who flew out for the rehearsals to help us.

Nothing can prepare you for the reality of the ODAC torture. Our presentation, given by one of our key investigators, went pretty smoothly, but then the comments from the advisory committee began. This committee is different for every NDA application. The FDA draws them from various groups: practitioners, academics, lay people with specific interests. This part of the meeting was awful. The worst part was the FDA rule that during the committee discussion the company cannot intervene. Some of the people on the committee were ill-informed, and others seemed very biased against the therapy. They argued back and forth. I remember looking across the room and seeing Christine Charet, the analyst from Nesbitt-Burns who had followed QLT and supported us for years. Her face was the color of a ripe tomato. I knew I must be flushed too; my skin was burning and my heart was beating so hard in my ears that I thought people around me might hear it.

The discussion at one point got so negative I was sure we were done for. "What a waste," I thought, "to be brought down by a few people who don't understand what we've done to try to bring a new therapy for people suffering from cancer." Some of the committee members started making theoretical suggestions about how we might construct a better clinical trial. They had to be academics who had no understanding of how difficult and expensive our present trials had been, and how long it took to complete them. I wanted to leave the room. I saw Alex get up and leave the room. I thought she might be going to throw up. I felt like I was going to. But she came back looking okay. Later she told me her nervousness made her want to pee so badly she had to leave.

There were a few moments of redemption. The advocacy groups made good arguments. And one of the doctors on the committee made a strong statement that he wanted this treatment in his armamentarium. It had a place in cancer treatment.

The committee voted on each indication up for approval. There were twelve people on the committee. I don't remember the individual votes but I do recall them being mixed. Some people voted against approval altogether and others voted for some indications and not others.

When the dust settled, we got approval for both the esophageal indications. The trial Ed had suggested and that Stu Marcus had scorned was approved. We had treated only fourteen patients with completely blocked esophagi, and Bob Temple was as good as his word – we got that approved as one pivotal study. We also got an approval for treating early lung cancer in patients deemed poor candidates for surgery. What we didn't get was the larger lung indication for local recurrence of tumors causing obstructive disease. Alex had been correct in doubting the data was strong enough to warrant an approval. And she had done as much as anyone could to clean it up. The FDA said we had do go back and try to recover data from the missing time points, largely from European sites.

Alex had worked so hard on the lung cancer file, she wasn't about to give up on it. The FDA had been fairly clear as to where the deficiencies in the filing were. There were holes in the one-week and one-month follow-up data. They asked for a reanalysis of the data assessing the length of time it took after treatment to achieve palliation. Alex was determined to get the information needed. It would take time, but she would get there.

CHAPTER EIGHTEEN

Taking Big Steps in Our AMD Clinical Trials

The Proctor and Gamble talks got bogged down in confusion over details. The instincts that many of us had had at the outset proved to be correct. After several months of discussion, Proctor and Gamble politely withdrew. I was relieved.

The situation with Randal did not improve. As chief scientific officer, I had little to do with him and he rarely gave me any directives; most of us were free to pursue our work and goals without interference.

Towards the end of 1995 we became aware that Randal was absent more frequently than he had been, without explanation. From the scant information we had, he seemed to be spending a great deal of time in Toronto. Ken, our chief financial officer, who shared an administrative assistant, Sandy Young, with Randal, saw a document carelessly left by Randal on a fax machine Sandy used, showing that Randal was applying for the position of CEO of Merieux-Pasteur, a French vaccine company that had taken over Connaught Laboratories in Toronto.

Connaught had been the only biopharmaceutical company whose head office was in Canada. Its origins dated back to the early twentieth century, when it became a producer of anti-toxins. After Banting and Best's discovery of insulin in the early twentieth century, Connaught had become a producer of animal-derived insulin. By the nineties it was basically a vaccine company. Under French management, it continued to produce vaccines.

If Randal were to act precipitously, springing his departure on us, the company would be damaged. We would have no obvious successor and would look like we were in complete disarray. Our credibility was shaky enough already. Only Randal knew what he would say in a press release and I was convinced by now he might not realize how much QLT might be damaged in the process. I feared his press release would paint him in a favorable light, with insufficient concern about how it made the company look. We had to stay one step ahead. We kept Duff Scott, the board chairman, informed as to what was happening.

Sandy let Ken know when Randal actually received the job offer and accepted it, subject to agreement on salary. We went into action and made a plan. We kept Duff Scott informed of what we thought Randal was doing.

We had a regular management meeting scheduled for a day or so after Randal agreed to his offer from Merieux-Pasteur. We figured that was when Randal would inform us about his imminent departure. Indeed, Randal informed us that he was sorry to have to tell us he was leaving.

While Randal was still making his farewell speech, Ken stood up and went out of the room. He came back with Duff, who had flown in from Toronto. Duff handed Randal a sheet of paper with the press release we intended to put out, announcing Randal's resignation. He told Randal that I would be taking over immediately as acting CEO and that Randal's services in the company were no longer required.

Randal was gone by the end of November, 1995. The news did not affect the price of our stock negatively. I was known to the investment community by then and was reasonably well regarded. My egghead reputation had more or less disappeared after my frequent attendance at investment banking meetings. I had been on the road with Bud and Ken many times and was on good terms with the investment community.

In January, the board asked me to stay on as permanent CEO. QLT was a different company than it had been in 1992. And I was a different person. Thanks to Bud, we had a solid management team in place. Before his departure he had made certain we had good people in key roles. I felt confident we could cover everything we needed to run a company efficiently. And I had learned a great deal in the previous five or six years. I was far more comfortable with the investment community. Ken was a brilliant chief financial officer and very effective with the public. David Main was an excellent communicator with investors. We could face Wall Street together. A search for a new CEO would be distracting and could be a disaster. We had not had good experiences with searches for CEOs. I knew where the company was going and what I had to do was make sure we got everything right.

On the other hand, I never wanted to be a CEO. I had enjoyed the role of chief scientific officer very much. I have never enjoyed being someone's boss, because I am non-directive. I saw my role in my lab more as mentor or guide. I decided I would continue in that role as CEO. The senior staff I had around me knew their jobs. I have always hated being told what to do, so I didn't see why I should start telling other people what to do, either. I wanted our management team to operate by consensus or as close to that as we could get. To avoid any possibility of nepotism, we had Ed report to Ken for purposes of performance review. He and Ken worked closely together in deal negotiations anyway. In all other aspects of his work, he reported to me.

Ed and I had no difficulties in working together, with me as his "boss." I don't remember any sensitivities arising. Ed has always been my biggest supporter. His ego is healthy and he never felt threatened by my position.

Ed and I make a habit of caucusing at bedtime. We both like bathing rather than showering. We take this time to talk over the day or whatever we feel like talking about. We never go to bed angry with each other. Over all these years, I don't recall a single difficulty around our relative positions at QLT. We sometimes had heated discussions about one issue or another, but never about the reporting situation.

The next year, 1996, started on a good note. Our results from our phase I/II ophthalmology trial with BPD continued to look very promising. Our partner was as excited as we were. Our shareholders rewarded us by pushing our share price up. We went from between six and seven dollars per share to numbers in the mid-teens. We began to get invited to present ourselves at significant American investment bank meetings.

It is rare when a company like ours has the kind of opportunity we had with macular degeneration. We had no competition for our therapy. Patients developing the condition had no recourse but to face the misery of losing their central vision. The possibility that we could provide a therapy that would save this vision loss was a thrilling and inspiring thought.

Analysis of the dose-ranging study we had completed showed that the treatment was both safe and effective over a fairly broad range of light dosing at a constant drug dose. The treatment regimen we finally chose involved injection of the drug, followed fifteen minutes later by a dose of light that lasted just ninety seconds. And that was it. It was quick, easy, and appeared safe.

With our partners, we decided that our phase I/II studies were convincing enough that we could go directly to phase III. This was a big risk. We had a new chemical entity. We should perhaps have selected one or two of the promising regimens we had seen in our first study and do a larger cohort of patients, following them longer. But doing a more extensive phase II would add at least two years to our development time. Alex, our regulatory vice president, rightfully took the conservative position in arguing for an additional trial. Andrew Strong, our clinical leader, felt we didn't need to do this – we knew enough to proceed. Our Ciba partners, who were inexperienced in drug development, agreed to take the risk.

The agreement with Ciba Vision had called for QLT to turn over clinical leadership to Ciba when we got to phase III. However, when the time came it turned out that QLT had far more experience in designing and running large phase III trials. So we proceeded to plan our big pivotal trials, of course in consultation with the joint project team. Andrew was in charge of putting the protocol together. He did a masterful job in finding our principle investigator and working with him to develop the protocol. Neil Bressler was a professor of ophthalmology at Johns Hopkins University. He and his wife were both retinal specialists. Neil had the respect of his peers and was a very experienced phase III trialist. And he was enthusiastic about our therapy.

We would run two identical trials at multiple sites in Europe and North America. It would be a double-blind study with a two to one ratio of treated to placebo patients. In pivotal trials like this, the sponsor has to propose a primary end point to the FDA. If the trial fails to achieve significance versus placebo in the primary end point, the trial is considered a failure even if named secondary end points are reached. We chose as our primary endpoint the percentage of patients losing less than two lines of vision at two years after treatment in comparison to the percentage of placebo patients losing the same amount of vision. "Lines" refers to the standard eye-testing chart and is self-explanatory.

Our secondary end points included a number of other vision measurements. Patients would return to their ophthalmologists quarterly. If their disease had recurred, they would be re-treated. They would be followed for two years.

We were faced with an ethical question as we prepared for the trial. We were providing a treatment for a disease for which there was no alternative therapy. What would we do if people asked to be treated outside the trial? The FDA permits treatments of this kind under what is called a compassionate use program or a special patient exception. What would we do if a family member developed the disease? Would we find patients willing to take the 33 per cent chance that they wouldn't receive treatment even if they participated in the trial? I thought of my mother, now legally blind, who would not benefit from the treatment now. What would she have done if she could have participated? I knew the answer. She would have participated with enthusiasm and hoped for the best.

One thing we knew: we couldn't be wishy-washy about providing treatment for some and not for others. The downside of a compassionate use program is that the company is compelled to keep definitive records of each patient treated and submit the records at the time of filing for approval. With the possibility that we might end up with hundreds of requests, this was a daunting thought. Neil Bressler, our principle investigator, who had had extensive experience with phase III trials, advised strongly against special patient exceptions, saying that our most ethical and best service to patients was to focus on getting the trial done as quickly as we could so as to get the drug out on the market and thus to patients as soon as possible.

We consulted medical ethicists who made the point that if we treated a single patient off protocol we would be honor-bound to treat anyone who requested treatment. I thought about my mother, whose condition at this time was too far along to qualify for treatment. Could I have been strong enough to deny her treatment if she were a candidate? I couldn't answer that. But we decided we would not ask for special patient exceptions and stuck to it, in spite of some extraordinary requests that turned into veiled threats and demands from several people of influence and in high positions who had developed macular degeneration. We had pressure from our Ciba partners when they got a request from a Saudi prince. But we stuck to the principled position.

We had our meeting with the ophthalmic division of the FDA and got agreement for our protocol. We had no difficulty in finding sites for the trial. Retinal specialists were interested and willing to participate. Physicians are not happy to have to tell patients they cannot offer them anything to treat a disorder, especially something as terrible as losing one's vision. The only thing these specialists could offer patients with macular degeneration was a possible treatment with a thermal laser that would essentially cauterize the leaky blood vessels. This treatment didn't always work, and patients usually irreversibly lost at least two lines of vision as a result of the treatment. It was certainly odd to be using a treatment that in effect caused the sort of problem it was trying to treat. Pure laser ablation was an option rarely used. PDT looked very promising in comparison.

We supported over twenty sites for the study and started recruiting patients immediately. We accepted all patients with wet AMD as well as patients with pathological myopia, a condition very similar to macular degeneration that occurs in severely myopic (near-sighted) individuals, usually much younger than macular degeneration patients.

We suddenly discovered we had competition from two small California-based PDT companies, PDT Inc. and Pharmacyclics. These companies had been in existence as long as QLT. PDT Inc. was founded by a former colleague of Tom Dougherty's, and Pharmacyclics by a former colleague of David Dolphin's. Both companies, like ours, had started out treating various kinds of cancer, but within a year of our announcing our results from our phase I/II trials both had jumped on the opportunity and were running studies for macular degeneration. I wasn't worried. We were ahead of them. I knew both their products and felt very confident that BPD had superior qualities to their materials.

There were three reasons for my confidence. First, my lab had expertise in biological systems. We had researched our molecule thoroughly; we understood how it distributed in blood and in tissues. We knew how it was processed in the body. Pharmacyclics and PDT Inc. were founded by chemists who were not that familiar with biological systems. They knew the photochemical properties of their molecules and had done the required amount of toxicology to get into human trials. But they hadn't done the in-depth research that we had.

Second, we had paid attention to developing a good formulation for our drug, partly at the behest of our partner, ACY. Now we had a robust liposomal formulation that was easy to handle.

The PDT Inc. formulation was something called cremophor, an oil-based product mainly used in animal research to test drugs that are difficult to dissolve. It is hard to handle and only used occasionally in pharmaceutical products. PDT Inc.'s product required a lengthy infusion cycle.

The Pharmacyclics product was water-soluble, which told me the drug was unlikely to accumulate with much selectivity in the vessel walls of the faulty blood vessels in the eye.

The third advantage we had was our lasers. We had agreements with two laser companies to produce elegant little diode lasers, no bigger than a compact photocopier, which emitted light only at 692 nm, the wavelength that activated our drug. Our competitors had not made similar arrangements and their drugs were activated at a different wavelength.

In 1996, we regained the interest of the investment community, largely based on our agreement with Ciba and the ocular indication. Analysts had looked at the numbers of cases of macular degeneration and the fact that there was little competition. Our early results had shown a proof of concept, so even though our phase III had risk associated with it, the odds looked pretty good to investors. Our stock price continued to perform strongly.

CHAPTER NINETEEN

An Open and Respectful Relationship

By 1997 our partnership with Ciba Vision was progressing well and we were optimistic about the future of BPD. Early in that year we succeeded in finding a partner to market Photofrin in Europe. Beaufour Ipsen was a privately owned mid-sized French pharmaceutical company. With Beaufour we now had in place all the partnerships we were seeking. We had Beaufour for European marketing, the US subsidiary of Sanofi for marketing Photofrin in the United States, and Ciba Vision for co-development and marketing of our AMD product worldwide. What we discovered during the next year was that management of these kinds of partnerships involves skill sets and diplomatic behavior that we had to learn as we went along.

Every company has its own corporate culture, nuanced by nationality. And every individual on any joint management committee has his or her own agenda. And that agenda is not always to the benefit of the collaboration.

I found that human interactions were rarely smooth, and it seemed that we in senior management at QLT were constantly trying to solve problems that ran from trivial differences of opinion to serious disagreements that had the potential to derail ongoing projects.

The reality was that we were all trying to do too much. We were still a small company, and although we had the expertise needed to cover all aspects of drug development, we didn't have nearly enough people with those capabilities. At that time we had barely a hundred employees, and most of them were researchers or clinical research associates. So our senior management ended up sitting on all the joint management teams we had with our new partners. In addition, many of us in management also attended major international scientific meetings that covered either photobiology or ophthalmology. That meant we all spent a great deal of time away from QLT.

Our newest partner, Beaufour Ipsen, was founded by Dr. Henri Beaufour in 1929. The company was still owned by the Beaufour family. The son of the founder, also Dr. Henri Beaufour, was now an elderly man. He was chairman of Beaufour's board, and had fallen in love with the idea of PDT. He had retired from running the business but was still very involved and influential with management. It was he who had driven the partnership through.

Our Beaufour agreement covered two areas. Their marketing group would market Photofrin in Europe. Other employees from their research facility outside of Boston would collaborate with QLT in a variety of research and development projects for expanding the Photofrin franchise by doing further clinical trials for other indications. There was also the possibility that they would acquire rights to our second-generation product, BPD, for use in treating cancer.

Beaufour Ipsen was unlike any pharmaceutical company I had come across before, although it probably resembles a number of private European pharmaceutical companies. It was large, having several thousand employees, though far smaller than the enormous recognized multinational pharmaceutical companies, which have tens of thousands employees. Its shareholder base was very small, because the Beaufour family owned the majority of shares. Their products included gingko, evening primrose, and a number of other natural products. The company was geographically dispersed, since they had a manufacturing plant in China, a head office in Paris, and a significant presence in London and the United States, where its flagship protein engineering research facility was located near Boston. The founder, Dr. Beaufour, had started the company driven by a desire to foster medical research and bring natural products to market for treatment of serious diseases. His son, it seemed, continued to build upon his father's dream.

Shortly after we signed our agreement with Beaufour, we were invited to a celebratory dinner with Dr. Beaufour at his elegant home in the Kensington borough of London. We had cocktails in his beautiful small London garden with its herbaceous border and paving-stone patio. Wines were provided from the family vineyards. He was a delightful gentleman who told me his wife had died of cancer after dealing with the debilitating effects of chemotherapy and radiation. He said he had been looking for innovative approaches to cancer therapy, and he believed PDT could be such a technology.

Our first business meeting took place in the Paris head office, where we met several of the senior management of the company. To my surprise, I recognized someone I had known years before. In the seventies, John North had been a post-doctoral fellow in the lab of a friend and colleague of mine at Berkeley near San Francisco. He had come to Vancouver and given a seminar at my invitation at that time, but since then we'd lost touch. I hardly recognized him, since he'd been long-haired and heavily bearded when he had been a fellow at Berkeley. Now he was smooth-shaven and short-haired, looking every bit the corporate manager. He worked at the London branch of the company in a kind of R & D coordinator position. He had been assigned to be the Beaufour coordinator for the projected research collaboration with us.

The Beaufour marketing people were located mainly in the Paris headquarters but were also spread out over Europe. Our contact person was located in Paris and was one of their senior people in business development, André Archimbaud, with whom Ed and Ken structured our working agreement. We all liked André. He had a good sense of humor and was very open. He was a widower bringing up a daughter on his own. His parents had emigrated from Algiers to Ottawa, so he was very familiar with Canada. He enjoyed gossip and provided us with quite colorful descriptions of Beaufour's past history.

Our collaborative research activities were orchestrated by their head of research, another colorful character called Jean-Pierre Moreau, or J.P. He was a biochemist and deeply involved with the activities of the sixty or so people he had in his R & D facility near Boston.

So we had two very different kinds of relationships that we had to deal with at Beaufour, one involving their marketing people and the other one with their research people. Lee Anne was pretty happy with the marketing group. They were professional, enthusiastic, and easy to work with. They were interested in expanding the Photofrin franchise, as were we. We already had a possible disease indication for that purpose and suggested it to Beaufour. They agreed with enthusiasm to our suggestion that we would try to expand the Photofrin market by running an additional trial for Barrett's esophagus, which was a condition on the rise in the nineties.

Cancer of the esophagus is a relatively rare malignancy. Until the eighties, the demographic for this cancer was mainly men at the lower end of the socioeconomic scale who were heavy smokers and drinkers. Also, the cancer itself usually developed in the upper segment of the esophagus, near the airways. In the eighties, the incidence of esophageal cancer started to rise and physicians saw an increase in esophageal cancers occurring in the lower end of the esophagus, near the gastric sphincter. Also, the demographic had changed and the cancers were now occurring, still mainly in men, but in people who were high achievers whose level of stress in the workplace was high.

The reason for this was the rise of a condition called Barrett's esophagus, which is a consequence of persistent acid reflux. The acid in the stomach is very strong and the stomach lining is designed to withstand this ferocious environment. The gastric sphincter normally protects the esophagus from this acid, but when people have chronic heartburn or acid reflux the lower part of the esophagus suffers injury from exposure to acid that escapes the stomach, and will undergo a kind of transformation. The damaged cells divide and start looking more like stomach cells than esophageal cells. The change in appearance is dramatic. Esophageal cells are pale pink while the cells lining the stomach are a deep red, so diagnosis of Barrett's esophagus is easy. These stomach-like cells are abnormal and prone to undergoing malignant transformation. So Barrett's esophagus can be considered a time bomb for esophageal cancer and is termed a pre-cancerous condition.

There were no suitable treatments for Barrett's esophagus. Antacids like Omeprazole were recommended, and might prevent worsening of the condition, but did nothing to reverse it. When malignant cells started showing up in the altered esophageal tissue, the only recourse for patients was esophagectomy, a ghastly procedure in which the esophagus is removed and the stomach is essentially hooked up to the back of the throat. Patients after esophagectomy will never enjoy a full normal meal again.

The use of PDT to treat Barrett's esophagus seemed to me an ideal application of the technology. The treatment would be similar to that developed for patients with full-blown esophageal cancer. It would eliminate the altered tissue and allow normal esophageal tissue to grow back. When I first became aware of PDT, I saw that there would be a problem using the technology where it would be most beneficial – in treating early disease as an alternative to surgery. It would be quick and non-invasive, but I soon realized that such a treatment would never be considered acceptable for early cancers. This is because cancer is a terrifying disease. Treating cancer is not like treating a benign blemish by burning it off. Physicians rightly want to be very sure, when they remove a primary tumor, that they have removed every cancer cell. Surgical removal of a cancer normally involves doing a wide excision around the malignant area to ensure that this is the case. This involves careful pathological examination of excised tissues. But after PDT, the area where the tumor was will be filled with inflammatory cells that are part of the healing process. Examination of that kind of area would never provide the desired evidence of complete tumor removal.

However, Barrett's esophagus is classified as pre-cancerous, and therefore the same stringent guidelines would not be deemed necessary. If patches of Barrett's tissue remained after treatment with PDT, the patient could be treated again. I was excited about the potential of Barrett's esophagus being an indication where a significant patient benefit could be achieved. Beaufour Ipsen was also excited by this possibility and agreed to collaborate on the study in Europe. We would work with J.P.'s group on this as well as deciding what disease condition we would select to start clinical development of BPD as the second-generation product.

Lee Anne became the de facto project leader for QLT in our dealings with Beaufour, coordinating with John North to set up meetings and arrange the complicated agendas. She was detail-oriented and extremely well-organized. And she enjoyed taking control of meetings.

We were very short-handed in both our clinical and regulatory departments. In clinical, Andrew Strong was completely swamped with the AMD program with Ciba. Alex, who had been leading the

Photofrin program, was overworked and behind schedule on getting the lung cancer filing in to the FDA. We had been looking for a medical person for some time and finally came across someone who appeared to be an ideal candidate, a man named Mohammad Azab. He was an employee of Sanofi in Paris, was a qualified oncologist, and spoke fluent French. He had come from Egypt originally but had studied in France. Our senior management was all in London for a meeting with Beaufour when we arranged to meet with Mohammad. We all liked him and thought he would be an important addition to our team. He would become the chief medical officer and vice president of clinical. Alex would remain the vice president of regulatory affairs.

On the investor front, David Main, our investor relations manager, had done a great job in getting fund managers to buy our stock. We were now being followed by most of the biotech analysts associated with the major Canadian investment banks. A majority of the reports written were positive and recommended investment in the company. Getting analyst coverage is a very important step in growing a biotech company. These analysts are called buy-side biotech analysts. They usually have post-secondary degrees in science or medicine as well as MBAs. Their job is to research the industry, looking for investment opportunities, and then recommend their chosen companies to bankers and portfolio managers. Analysts are free to choose to follow a company or not. If an analyst decides to follow a company, they produce an in-depth report on that company, the progress it is making, and the risks the investor is exposed to. Then they recommend to the investor to buy, sell, or hold a given stock. They update their evaluations at regular intervals.

Most of the analysts covering us recommended "buy" for our stock, which had risen to between $15 and $20 per share. The shares had been at between $6 and $7 dollars when I became CEO. None of the investment banks in the United States paid any attention to us yet. People were investing in the promise of the AMD treatment, and we wouldn't have the results of our phase III trials until late 1998 or early 1999 – more than a year away – so any appreciation of share value during this time was driven by positive market results from Photofrin. No one expected Photofrin to be a blockbuster, so expectations for it were realistic. There are many ways to treat late-stage cancers. Often patients are ill and frail and would not be good candidates for an intervention such as PDT.

In late 1997, the FDA responded to the filing for the Photofrin treatment of progressive lung cancer and a date was set for a meeting with an advisory committee. It was during our preparation for the advisory committee that we ran into problems. Alex was the head of regulatory affairs. Until we hired Mohammad Azab, she had been handling virtually all clinical matters pertaining to Photofrin filings as well. Now Mohammad, as chief medical officer and an experienced oncologist, became part of the mix. He wanted to leave his imprint on the process, whereas Alex had done all the preparatory work and knew the file inside out.

The preparatory rehearsals for the advisory committee with the FDA became toxic for everyone involved. Alex and Mohammad bickered incessantly over details. I could see both sides and worried that the company would suffer at the hands of the FDA as a result of these internal disagreements. Eventually we were successful with the FDA, so we were able to live up to the promise we had made to Sanofi US that we would get the approval for late-stage lung cancer.

But the damage had been done, and Alex and Mohammad were never able to work harmoniously together after that. They tried, but the antipathy ran deep, particularly on the part of Alex who had felt insulted and hurt. This was a serious loss. Regulatory affairs and clinical research are very closely linked and one wants the two units to work seamlessly in concert.

We had the Barrett's esophagus trial to design (again, a promise we had made to our partners that we were under obligation to fulfill). Mohammad came up with a simple design that seemed too good to be feasible. We would compare our treatment against standard therapy. The end point would be reduction in the amount of abnormal tissue remaining in the esophagus after a course of treatment. Standard therapy was Omeprazole, an antacid pill, which in some cases does delay increase in Barrett's tissue in the esophagus but does not usually reduce it. Our treatment, on the other hand, was designed to destroy Barrett's tissue.

Alex did not think we had a chance of getting the okay from the FDA to use this as a registration trial. However, the FDA proved her wrong and gave us the nod to run this study as a registration trial.

This outcome did nothing to improve the relationship between Alex and Mohammad.

Meanwhile, we also had to manage our external relationships. Our regular meetings with our Beaufour and Ciba partners took place in Europe more often than in North America. We found that the management of partnerships can be fraught with unexpected difficulties. Individuals develop specific likes and dislikes that can poison relationships. Our Ciba partnership, I think, ran as smoothly as any complex relationship can. We had designated Gusti Huber as our project manager and chief liaison between Ciba and QLT. Ed was designated the liaison person for QLT. Ed and Gusti developed an open and mutually respectful relationship that was critical to the project running smoothly. Gusti was an honorable man who believed completely in the project and also in the expertise that QLT had developed. He knew the weaknesses inherent in his own organization and always kept the interest of the project itself at the forefront of any actions.

In late 1996, Ciba-Geigy merged with Sandoz, another Basel-based Swiss pharmaceutical giant, to form Novartis. This merger had essentially no effect on Ciba Vision, which continued to operate as an independent entity.

I was impressed by Swiss cultural mores. Depending on where you are in Switzerland, there will be a cultural influence of either French (in Lausanne), German, or Italian (in Lugano) origin. Zurich, Basel, and Bulach are located in the northwest part of the country and are close to the French and German borders, so one feels the influence of both cultures. In spite of these geographic differences, Switzerland has its own unique character. It is the tidiest and cleanest country I have ever been in. Everything looks carefully ordered. This order works its way into the behavior of the Swiss. Where we might respond to a question about how long it would take to get from one place to another by rounding up or down to an approximation like "about half an hour," in Switzerland answers will come back with precision – "twenty-seven minutes."

I noticed this exactness in the way our meetings unfolded with Ciba Vision. Agendas were concise and accurate, discussions were to the point and I had the impression that there was no hidden agenda. Our teams worked productively together.

Our meetings with Beaufour Ipsen were a stark contrast to the ones with our Swiss partners. Meetings in Paris did not happen too often but when they did, their marketing manager and André Archimbaud, our business development point person, made sure they took us to first-class restaurants. Lunches could take up to three hours and were taken very seriously. No one seemed in a hurry to get closure on any points of discussion. But progress was being made, regardless of the apparent lack of urgency.

Marketing in Europe is complicated because each country handles its marketing strategy somewhat differently. I had no way of evaluating their competencies, but I trusted Lee Anne's assessment. She was a perfectionist with extremely high standards, and she seemed satisfied with the way the Photofrin launch was being handled. The corporate culture at Beaufour was relaxed. The senior managers, it seemed, had a great deal of freedom. The CEO was semi-retired, the company was in a solid financial state, and its established products enjoyed sizeable markets.

When we had meetings in Paris we always tried to add a couple of days of vacation time on to our schedules. Lee Anne and I would enjoy afternoons of shopping in the boutiques and then find a bistro and eat *canard magret* and salad with wine. On another occasion, Lee Anne, Ed, and I got tickets to the French Open tennis championships at Roland-Garros and saw Steffi Graf playing against Aranxia Sanchez Vicario in a semi-final during which the audience booed Sanchez Vicario for the moon balls she lobbed.

When our meetings were organized in other locations, we had less fun. The London offices were quite far out of the West End of London, west of Holland Park near Olympia. I never knew what exactly went on at the London office, but John North, our contact person, had his office there. The atmosphere in the London office was very different from that in Paris. It was not relaxed. There seemed to be a level of anxiety and confusion there. I asked André about this at one point and he told us that there was indeed uncertainty in London and worry that the offices might be closed.

I came to believe that J.P. was uninterested in any joint research project with us. He was obsessed with his own research. I don't blame him for that. He was passionate about what they were doing in the Boston lab. But I felt he could have given us the courtesy of making his lack of interest known and delegating the responsibility to someone else. Our discussions of how to run the Barrett's esophagus trial in Europe flagged. Various team members introduced artificial barriers.

On the other hand, John North had become a strong believer in PDT and QLT. He was intelligent – a scientist who was well-grounded in basic science but had had experience over the years in drug development. I was glad we had him in our camp. I wondered if he would become our Gusti at Beaufour. Gusti was able to work within his Ciba system to get things accomplished, even when there was opposition.

QLT now looked like a company that knew where it was going. We had achieved a lot. We had a drug approved in Europe and North America that was being marketed by legitimate pharmaceutical partners. And we had a pharma partner for our second-generation product. We just had to keep everything moving along until we finished our phase III clinical trial in AMD.

CHAPTER TWENTY

Our Work on Age-Related Macular Degeneration Is Her Legacy

Being CEO didn't change my already heavy workload in any significant way. We were all extremely busy. The atmosphere around the company was positive now. We knew where we were going. I felt the company had a stability I hadn't sensed before. We were excited about the AMD trials, which were progressing at a rapid rate. Doctors were keen to offer their patients a possible treatment for a devastating condition. Unfortunately, it was too late for my mother. She had lost her central vision by now and she was becoming very frail. I felt awkward and somehow guilty talking to her about the work at QLT.

I felt happy that we'd played a part in bringing PDT to market as a possibility for patients to receive another form of therapy for various kinds of cancer. Existing treatments for cancer could be devastating to the patient. Radical surgical procedures can leave patients permanently damaged, and both chemotherapy and radiation therapy can ravage the body. With patients who have a high probability of recurrence and possibly a short life span, quality of life issues become significant. PDT wasn't going to solve all cancer problems, but it did provide a non-invasive and benign procedure for local control of cancers. At QLT, we were believers in the importance of this new option for patients.

I was so busy that I was grateful that our children were more or less independent by this time. Jennifer had finished law school and her articles at New York University and had met the man she would marry, Paul Schwalb. She seemed very happy. She was employed at Brooklyn Legal Services helping disadvantaged people who were in danger of losing their homes. I got to see them quite frequently because we were very often in New York, where we had a large shareholder base.

She and Paul got married quietly just before Christmas that year. Ben had three children by then and was running his own restaurant. I was still heavily involved with the research going on in the company and in my lab at the university. That may have been the reason I failed to pay much attention when my assistant, Gail Rothery, developed huskiness in her voice that persisted through the summer. She joked about it, saying she liked how sexy she sounded. She was a smoker. One day she showed me an unexpected bulge in her neck. She tipped her head to one side and extended her neck. When she did that, it was easy to see a distinct lump about three centimeters long. I put my hand out and touched it. It was hard and a little smaller than a golf ball. The lump had to be an enlarged lymph node, and that meant Gail had either a serious infection in her larynx or something much worse. I asked her how long it had been there and she said she wasn't sure when she first noticed it. It was growing. It was bigger now. I asked her if it was sore. She said, "Not really." I told her to get a doctor's appointment right away and have her throat examined.

Gail and I were friends. We both liked antiques and auctions, so frequently Gail, Ed, and I would go to Maynards antique auctions and then have a late dinner out. Gail lived alone and didn't drive. She was unmarried but by her own accounts had not led a nun-like existence. She was definitely a party girl who liked red wine, cigarettes, and a good time. She had been the lover of a quasi-gangster once. She had had many lovers, but she told me that one morning about ten years earlier she had woken up, looked at her current boyfriend asleep beside her, and decided she didn't want men in her life any more. And that was that, she never looked back.

Gail was diagnosed with laryngeal cancer. I accompanied her to her first appointment at the British Columbia Cancer Control Agency. I was jittery. We sat in silence throughout the surgeon's outline of the tumor in Gail's throat and the recommended treatment. Gail sat silent, still as a statue while this was going on. She would have to undergo radical surgery, a complete laryngectomy plus regional lymph node excision.

She would lose the ability to speak normally. Her larynx (voice box) would be gone. There would be a device implanted in her throat with an aperture in her neck through which she could inhale and exhale. She would be able to speak after she learned how to use the device. It was a matter of learning how to make the sounds by expelling breath through the hole. He showed us a video of a patient following the surgery.

I had been around people who had had this surgery and found their speech virtually incomprehensible and freakish. The video didn't sugar-coat the outcome.

I saw Gail's face take on a steely look as we sat watching. Afterwards, she continued to sit silently. Then she said, "I'm not going to do that," in a decisive tone.

The surgeon looked surprised. Gail looked at me. "I won't," she said. "I won't go through life making sounds like that. I wouldn't be me anymore. I'd rather die."

I knew she meant what she said. The surgeon recovered his aplomb. If his patient was going to refuse surgery, she'd be referred to a radiologist for radiation therapy. Gail agreed to see the radiologist.

Gail was enrolled for radiation therapy. She had no immediate family in Vancouver. I'm not sure who at QLT organized the roster of people going with Gail for her radiation treatments, but she never went alone to the cancer agency. Someone was there to give her a ride and make sure she had food in her apartment.

I tried to be optimistic. Head and neck cancers do not commonly metastasize to distant organs, although many people will suffer from locally recurring disease. And many head and neck cancers are responsive to radiation therapy. Maybe Gail would be lucky. Her tumor was shrinking as a result of the radiation.

Soon after Gail completed the course of radiation therapy, I had to travel to Japan for a meeting with Ciba Japan on the ocular program. The Japanese wanted more toxicology work done to ensure safety, since they were concerned that Japanese eye shapes differed from Caucasian eyes. This dimensional difference could alter the effectiveness of the treatment.

We had enrolled Asian patients in our phase I/II trials and had seen no differences between them and Caucasian patients. We were including eye color as part of our patient database in our phase III and would eventually know if eye color would have an effect on the efficacy of our treatment. But we didn't know at this time. I knew it was going to be a tiresome meeting. In my many trips to Japan during those years, I found myself growing more confused about Japanese culture the more I was exposed to it.

The rules of conduct in Japan are rigorous and not intuitive to people from Western cultures. Whenever I went there I was always certain I'd screw up in different ways. I usually towered over my male counterparts and felt gawky as a result. I sensed how uncomfortable I made my hosts. Most of the people I met had never encountered a female CEO.

I was informed that gift exchanges between CEOs were mandatory. I had a meeting scheduled with the CEO of the firm that was formulating our liposomes and thought I had found an ingenious gift, a beautiful First Nations–designed silver letter opener.

When I presented it to the CEO, he gleefully opened the gift-wrapped box and then his face fell. He asked me what it was. To him it looked like a knife, and giving either scissors or knives as gifts, I found out later, is a complete taboo. Such a gift portends a breakup of any relationship. I said it was a letter opener but he looked only slightly less upset.

When I got back I was dismayed to hear that Gail was in the hospital. She had been admitted for abdominal pain. She had suffered acute nausea and had more or less collapsed, according to a neighbor who had taken her to the hospital. Lee Anne Pilson, our marketing VP, and I went to see her in the hospital. We found her in a private room hooked up to an IV. She told us she had some kind of blockage in her gut and was not allowed to eat or drink. She could suck ice cubes only. That could only mean that her gastrointestinal tract was completely blocked. She was dying. I think Lee Anne knew that too.

It didn't appear that Gail had acknowledged her condition. She said her stomach was tight and showed us by pulling up her hospital gown. Her abdominal area was distended. I didn't like the thoughts that intruded, telling me that we were in the end game with Gail. Her tumor had spread to her gut and was going to kill her.

The following morning, which was a Saturday, I received a call at about seven from one of Gail's oldest friends. Gail had had a very bad night and they had moved her to the step-down unit. A step-down unit in a hospital is a place where patients receive a level of care one step down from intensive care. Nursing staff is there at all times and most patients are terminally ill and sedated.

Gail was heavily sedated and unlikely to regain consciousness. If we wanted to say goodbye we should get there soon. I contacted Lee Anne and Sandy and let them know. Ed and I picked up Lee Anne on the way to the hospital.

It was still early in the morning but a couple of Gail's friends were there, people from her life before she came to QLT.

Gail was lying peacefully in the bed, although her breathing was somewhat labored. I noticed a plastic tube delivering a thick yellowish fluid into a collection bottle under the bed from under the sheet covering Gail. I realized she must have had tumor cells growing with abandon in her peritoneum and that the tube had been inserted to lessen the pressure. The thought made me cringe.

Word spread and people began trickling into the ward. There must have been at least a dozen of us standing around her bed by about ten that morning.

Then she opened her eyes. Her vision was clouded and she seemed confused at first. She gazed around and saw that she was now surrounded. She looked from one to another of us. Then she said, "Wowzy!" A strange word to be uttered from a deathbed.

I don't know how much morphine Gail had in her system, but whatever the amount it wasn't going to stop her from enjoying her last party. She rallied, she joked, she laughed, she rued the fact that she hadn't spent all the money in her bank account.

The day stretched out. The nurses, used to seeing people die, seemed mystified by her resilience. She didn't start flagging until late in the afternoon. Her voice became slurred and she wasn't completing her sentences. She died in the early evening after having been a wonderful hostess all day long.

She had been right not to undergo a laryngectomy. I agreed with her that such an operation would have taken away her soul. And given the fact that she had such a swift onset of metastatic disease, she would probably have died in the same time frame, having gone through debilitating and painful surgery. She had been a memorable personality at the company and she would be missed.

Gail's struggle brought home to me again how important it is for cancer patients to have options regarding their treatment. I thought about Merrill Beil, one of our PDT pioneer physicians. Merrill was a head and neck surgeon who specialized in treating patients who suffered from local recurrences of their head and neck cancers. Most of us at QLT had attended scientific meetings where Merrill had presented his data in the form of graphic slides of his patients' treatments. He took heroic surgical measures and then treated the tumor bed with PDT using Photofrin. He had had some remarkable outcomes, with patients surviving for years tumor-free. Gail would not have been eligible for Merrill's procedures because what he was doing was experimental, and patients have to fail conventional therapies before they can be treated with experimental procedures.

At the same time that Gail was fighting her cancer, I knew my mother's health was failing too. Three years earlier, we had tried to persuade her to let us buy a house with an independent suite in it for her. But she had resisted, saying she felt she wanted to maintain her independence and would move to assisted living. She was ninety-two and legally blind. Much as I would have liked to have her close to us, I knew not to argue with her. It was my anxiety about her living in her own home up to that point that had driven me to suggest she move in with us. I dreaded finding her lying at the bottom of her basement stairs, having missed a step. She'd done that before, when my father was still living, and had broken her collarbone. By now she was also quite deaf. I admired her fierce independence but felt enormous relief when she decided to move.

In that same summer that Gail got sick, my mother, who was by then ninety-four by then, was hospitalized because she collapsed in her room. She was very anemic and it was found that she was bleeding into her intestines. Blood work showed that her bone marrow was giving out. She was no longer able to produce the needed number of cells that would differentiate into platelets. Without platelets, her blood would not clot appropriately. While platelets were the first sign of bone marrow failure, it would lead to the gradual loss of the ability to produce the right number of red blood cells and other cellular blood components.

I consulted with her doctor. As I had expected, there was no cure for her condition. Perhaps today stem cell transplants would be a possibility. In 1996 treatment would involve monitoring and blood transfusions.

In the fall she had another major intestinal bleed and ended up in the hospital again. I visited her and she seemed quite cheerful. She had had a blood transfusion and felt better. The nurses said she would be released the following morning. I said I would pick her up and take her back to her apartment in the care facility.

When Ed and I arrived the next morning, I knew all was not well. My mother was deathly pale and lying in her bed. The nurse told me she had had another serious intestinal bleed and that her gut was perforated. She was dying.

We went to her and sat by her. She said how cold she was and that her chest hurt. I wrapped her in my arms and held her close for a time. She said how good that felt. I was so glad I could do something. She complained again of being in pain. My mother never complained of pain, so I knew it was bad. We called the nurse and suggested my mother be given something for her pain, and they brought an opiate of some sort that they gave to her orally, because the nurses didn't have the authority to access the injectable opiates. The doctor was expected any time. They'd give her an injection as soon as he got there.

My mother whispered to me, "Is this it?"

The opiates took effect and my mother lapsed into a peaceful sleep. She died very soon after that. Our son, Ben, walked in the door of her ward as she took her last breaths. He felt very grateful he'd arrived to be with her at that moment. I felt an inrush of gratitude and relief that she wasn't in pain any more. And then came the emptiness. I thought I was prepared for her death. After all, she was ninety-four. And I knew she was failing.

But I was not prepared. I doubt that one ever is. In the following weeks, she haunted me. I dreamed about her. I still do. I didn't grieve so much as feel her presence as well as her absence. I had a profound sense of loss. I thought more about what a powerful influence she had been on my life and development, without ever giving me any direct guidance. She had done it just by being who she was and letting me be who I was.

I think of our work on AMD as her legacy.

CHAPTER TWENTY-ONE

We at QLT Were on Pins and Needles

We finished recruitment of patients for our phase III trials for AMD in the fall of 1997. There were over twenty clinical sites in Canada, the United States, and Europe. We had enrolled close to seven hundred patients. Two thirds of those patients had received our drug and a third had received a placebo.

We had agreed with the FDA to evaluate the patients after a one-year follow-up. Since the last patient recruited was in September of 1997, analysis of the data would start in the fall of 1998, one year after the last patient was entered into the trial. The FDA had agreed that we could file for drug approval based on twelve-month results, provided that we follow patients for an additional year for safety purposes as well as to determine long-term effects of the treatment. But we could get approval based on the results obtained at one year.

In negotiating trial conditions with the FDA, the FDA insists that a company choose a single expected outcome as what is called the primary end point. The choice of a primary end point is a critical one in setting up registration trials. The company proposes an outcome that it believes can be reached and then gets approval from the FDA that the predicted outcome is a suitable assessment of efficacy. Failure to reach that primary end point is considered a failed trial, regardless of however many other beneficial results are achieved. Our primary end point, the one that would give us either a thumbs up or a thumbs down from the FDA, involved evaluating how much visual acuity the two populations would lose in one year. In setting up the trial, we predicted that there would be a significant difference between patients treated with our drug and those on placebo, who lost two lines of vision or more at one year. (A line of vision represents one line on a standard eye chart.)

The way the protocol worked was that the treating physician saw their patients every three months following the first treatment. If the leaky vessels persisted they would be treated again at that time, so it was possible that a patient could receive five treatments over the course of one year. The last patient recruited would come into the doctor's office for his/her one-year visit in the fall of 1998.

We had a number of secondary end points identified in the study. These included patients losing four lines of vision and loss in contrast sensitivity. Contrast sensitivity involves visual capability in distinguishing differences in shades of gray. This kind of visual acuity is important in reading things like newsprint.

The study was double-blinded. When a physician entered a patient in the trial, their name was submitted to a data manager who was in charge of randomization. The patient was assigned a number and randomized to either placebo or treatment. When the patient arrived for treatment, the study nurse provided the physician with a prepared syringe containing either drug or placebo, without his or her knowing what was in the syringe.

These kinds of trials are very expensive. They require several layers of administrative control to protect the randomness necessary in a double-blind study. In this trial, clinical investigators received $12,000 for every patient they enrolled. On top of that cost, the company paid for the additional staff required at each site. The enormous risk of phase III trials can be mitigated to some extent by running what are called phase IIB trials, where more clinical data are gathered about the drug. The difficulty with these is that they too are enormously expensive and time-consuming. Investors in biotech companies are patient, but even these investors will not reward a company for dragging its feet. Phase III trials are always a very expensive gamble but the risk/reward ratios are very tempting.

For all new blinded drug trials, the FDA also requires an oversight body called a data safety monitoring committee to monitor safety issues during the trial. This committee may have up to a dozen members, selected from specialists in the particular field involved as well as qualified biostatisticians. This committee does interim analyses of the incoming data in an unblinded manner. They are the only people who know how the trial is progressing. The reason for this additional oversight on new drug applications is mainly safety. If patients on the treatment arm have more serious adverse events than placebo-treated patients and the committee believes these events are likely to be treatment-related, it may decide to advise the sponsors that the trial should be shut down. Recommendations are shared with the FDA. Alternately, if the data are stunningly good they may also shut the study down and recommend accelerated FDA approval. Such an occurrence is rare, because the integrity of the completed study provides important scientific and clinical information, and the oversight committee may have safety concerns even when the data are compelling. Also, a well-controlled complete study provides important information the company can use for marketing purposes..

All through those fourteen months, we at QLT were on pins and needles. The only information we were entitled to see were the results of the visual acuity and other assessments that came from the physician's offices after a patient was seen. But we had no way of knowing whether the data came from a patient who had received the treatment or the placebo, because the data were blinded.

Andrew Strong, our key clinical person on the AMD trial, and I were both obsessed with these raw data, especially lines of vision lost by patients. What we were able to observe was that there was a wide spread in the visual acuity of patients returning for their quarterly visits to the doctor. Many patients were retaining the visual acuity they entered the trial with. When we saw that about a third of the patients treated were experiencing more vision loss than the others, we hoped that those were the patients on placebo, and that those doing well were on treatment. We had been told repeatedly that all patients diagnosed with AMD, without exception, lost significant vision within the first year of diagnosis, so we had reason to feel optimistic.

I was very impressed by the seriousness with which our clinical investigators took the integrity of double-blind studies. There are probably multiple means by which a physician can get an inkling of which of their patients have received treatment and which have received placebo, but I never got the faintest indication that our investigators did anything even slightly questionable. Good phase III trialists are a special breed of clinical investigator. They are rigorous in their record-keeping and in the way in which they follow a protocol. And they don't even consider trying to break any rules. They are quite different in mindset from those physicians who excel in phase I and II clinical studies. Phase I and II trialists usually don't conduct blinded studies, and they are interested in fine-tuning a treatment protocol to try to make the outcome better. In a way they are more scientific in their approach. Both types are critical to drug development.

Neil Bressler, our principal investigator, was a superb phase III trialist. He had led other major phase III trials run by the National Eye Institute. His rigor and clinical integrity were unquestioned, so his credibility was substantial.

Preparing the documentation for filing for drug approval is another gargantuan undertaking. When Alex had prepared the filings for Photofrin she had had the American Cyanamid infrastructure behind her, churning out statistical analyses and carrying out the quality control and quality assurance required. We had assumed that when the trial itself was over, Ciba would bring in phase III expertise and infrastructure from their parent company, Novartis, but it turned out that they had not considered that. We realized that Ciba Vision had no experience in preparing the extensive documentation required for filing with the FDA for approval of new injectable drugs. And certainly they had none of the infrastructure needed. Because of our Photofrin experience, we at QLT had far more expertise in those kinds of filings than they did. So it was agreed that, providing our data supported a filing, QLT would be the lead company in putting together the submission for the United States and Canada. Ciba Vision would use the information package that we generated to make the appropriate filings in Europe, with help from their head office. So we were in charge of data management. That meant we had to hire a lot of people in a very short period of time. Fortunately, we had found a very experienced person to head up our human resources department, Linda Lupini. At one point she and her staff were hiring one person a week.

In order to have a successful launch of a new product like ours onto the market, it is essential to have all aspects perfectly coordinated. If all went well, we should have our data cleaned and ready to announce by the start of 1999. It would probably take six months to prepare the US filing after that, with the European filing a few months later. That should mean we could expect approvals somewhere in the first quarter of 2000. Ciba marketing people wanted to be ready to launch the product immediately after an approval, literally within days. We made sure the companies that were making our lasers would have adequate numbers of instruments available. We also had to ensure there were adequate quantities of drug available for the launch. It was assumed there would be a rapid uptake of the product in the United States. All these preparations had to be started, and money had to be spent well ahead of the time when we would get the decision as to whether our drug was approved in a particular country. (Approvals are country by country, though in Europe there is also a non-binding EU-wide system as well. An approval in one country wasn't a guarantee that it would be approved in another, though every approval raised the odds.) This is a gamble that companies have to take in order to have a successful product launch.

We had to have a commercial name for our product. Naming a new product is a complex process. The suggested names have to be rigorously examined, and translated into many languages to determine if, in translation, the name can be offensive in some languages. "Safe" candidates for the name are then circulated to the sponsoring companies. One of the candidate names that got strong support was Amdyne. But then we heard our Atlanta colleagues at CIBA Vision pronouncing it, and it sounded like "I'm dying." We settled instead on the name Visudyne™.

We started to get more interest in QLT from US-based investors as the time approached when we would have our phase III data. Our stock traded on both the Toronto stock exchange and on the NASDAQ exchange in New York. The volume of trading started to increase and the price edged up, but not very much. Investors in biotech are similar in many ways to people who speculate in gold mines. Failures in phase III clinical trials happen upward of 50 per cent of the time. And if the trial fails, the stock price will plummet. The reason that investors take the chance before results are known is that after announcement of the results, stock prices usually rise sharply. I have talked to many investors over the years and asked them about how they make decisions on investment before results are known. The answers are surprisingly varied. CEO credibility is a major factor, along with successful track records of the drug and the management team. One portfolio manager said he based his investments on how easy it was for him to understand the science when the CEO explained it. I was pleased to learn that this man became a big investor in QLT.

Since we were responsible for preparing the filing for the FDA, we knew we had to hire a lot of people into our clinical and regulatory departments – statisticians, data managers, quality technicians, medical writers, and clinical research associates. We would also have to file separately for approval of the two lasers, so we had to increase the number of people in our devices division. Filing for a new drug approval (NDA) also requires an incredible amount of information on the drug itself.

This aspect of a filing is called the chemical manufacturing control (CMC). Our manufactured product was complex, involving a number of steps. Every step of the procedure has to be thoroughly documented with records of stability data and documentation of every batch of product made, along with toxicology reports. The FDA requires that when a company develops a procedure for manufacturing a product, at some point the process is "frozen"; that is, from that point on, the procedure cannot be fine-tuned or changed in any way. Minute levels of impurities have to be documented and identified and "specs" laid down for future batches of product. Any deviation from these specs results in a failed batch of product.

Manufacturing is a painstaking, very expensive and arduous process and requires the hiring of considerable numbers of quality control and quality assurance technicians. Many junior biotech companies have been unpleasantly surprised at the eleventh hour by manufacturing issues, and their products delayed from marketing approval while they clean up their manufacturing. Fortunately, we had cut our teeth on Photofrin, which was a manufacturing nightmare in that it was highly unstable, and we saw many failed batches because the drug didn't pass the specs set for it. So we were well-prepared and didn't anticipate any manufacturing issues. Unfortunately, we were proven wrong very late in our filing process.

Another sizeable document that had to be prepared for the filing was a detailed summary of all the scientific research that had been done on our drug, including the early clinical work on skin cancers. Our research staff took on the responsibility for putting together that document.

The anticipated day arrived when the last patient came in for her one-year follow-up. This was in September, 1998. It was an important day for us. Our fate was now sealed and in the hands of our clinical research assistants, who had to make sure our investigators adhered to good clinical practices for the next year. Clinical research assistants are trained to interface with the company and the investigators. They are either nurses or people with a medical or science background. Their job involves visiting clinical sites while trials are ongoing, checking that good clinical practice is being maintained, and generally monitoring the way a trial is going and the way data are being entered into individual case report forms. When the trial is complete, the research associate is responsible for collecting all the data from a given site. Each patient will have a pile of documented information, not just pertaining to their visual acuity and vital statistics, but also records showing any kind of problems they might have encountered during the time of trials, any health issue like headaches, nausea, and so on. Statistical analyses are run on the adverse events to determine if they are treatment related. In order for the data from each patient to be used in statistical analyses, the information has to be painstakingly transferred to the appropriate forms for entry into the analytical programs. Recently it has become possible to do these things electronically and from a distance.

The work carried out by clinical research associates then has to be checked by someone from the quality control department before it can be signed off. A company like ours would stand or fall on the outcome of this trial, so it was critical to make sure everything was done in a professional way. All these controls have to be in place to satisfy FDA rules. Agents from the FDA can come and visit the company at any time, especially if there are concerns about trial integrity. The FDA will, at some point, audit all companies that are preparing a filing for approval of a new drug. The audit will include visits to some of the clinical sites, as well as some of the locations conducting tests that make up the overall record. They require access to all databases in order to ascertain that good clinical and manufacturing processes are being followed. These audits are extremely nerve-wracking for companies such as ours.

Anxiety tempered with hope and excitement pervaded QLT during those months while the data were being entered. Everyone was working long hours, but our hopes were running high and you could feel the anticipation in the air.

Christmas approached and the pressure mounted. We hoped to be able to announce our results early in the new year – an ideal time, before the biggest biotech investor meeting began in the second week in January. In 1998 Hambrecht and Quist (H & Q), a major New York–based investment bank, sponsored this meeting, called H & Q. The meeting set the tone for investor picks of biotech companies for the year. I don't think any of us really thought our drug would fail. I know I didn't think seriously of the possibility, at least until the time I realized we'd soon know how the trial had gone. In a way I think our optimism was a sign of our naïveté. Phase III trials are fraught with opportunities to fail, and we had gone directly into phase III without doing an appropriate phase II trial. We had taken chances, but in our phase I work we had seen good indications of both safety and efficacy.

People started working late into the night. We set up a kind of cafeteria in an unused space in our building so that hot meals could be served to those working long hours. Everybody was in a state of barely controlled excitement.

Then we got to the data lock. This meant that all the entries were in the database; they'd been checked and were ready. It wouldn't be long now. Probably in two or three days we'd know the results of our trial and whether we had reached our primary end point. The statisticians had to go through a series of dummy runs to make sure the very complicated analytical system was working properly.

When we looked at the timelines for meeting with our partners and the data safety monitoring committee and releasing our data to the public, another serious problem reared its head. If a public company has knowledge of a material event (any event that may affect stock price), the company is obliged to make that information public as soon as it is known. Any leakage of information that could lead to insider trading is a serious issue. We had a significant confidentiality issue that we had to deal with, seeing that we'd know our results, as would a considerable number of employees, in the time interval between Christmas and the new year. Although we might know our top-line information, we wouldn't be able to announce our results until after we met with our Ciba partners and got the thumbs up from the data safety monitoring committee. We were scheduled to meet with our partners and the committee in Atlanta on December 29. There we would frame the press release and decide strategically how to position ourselves. We would put out the announcement on January 2. Whether the results were positive or negative, we had to make sure that no one in the company told anyone about our results before we went public with the information. Stock exchanges tend to monitor trading patterns during sensitive times like these. They can trace where trades come from and determine if any unusual trading is going on with insiders. We had impressed this on our employees, pointing out that insider trading is a crime. This was before the famous Martha Stewart fiasco in the early 2000s where she did precisely that: bought shares as a result of a tip given her by the CEO of a biotech company before results were made public.

Those three days of waiting for the results to come out dragged for all of us. We had been so busy, and now we were in limbo. I couldn't even think about Christmas shopping. I started having nightmares. I dreamt I was sitting in my office and saw Mohammad sneak up to my door and slip a message under it. When I picked it up it was a note saying the trial had failed. I told myself I had to be prepared for the worst. A failure would be a disaster for the company. Investors would leave en masse. We'd have to let dozens of people go. Would we survive? I wasn't sure.

CHAPTER TWENTY-TWO

People Couldn't Stop Smiling

Mohammad phoned me at home in the evening on December 23 to give me the results. Our primary end point had been achieved with highly significant differences between the treated and placebo groups ($p < 0.001$). Mohammad went on to say we'd also achieved significance for all our secondary end points. The treated and placebo groups were all highly significantly different in every category.

Ed and I opened a bottle of champagne and toasted the team, the drug, and QLT. We'd done something incredibly important. My overarching feeling at that time was enormous relief. We hadn't let people down. I wanted to hug everyone we'd worked with. And I felt very happy. I knew if our trial had failed, instead of relief I'd have felt a sinking feeling in my stomach. My whole body would have reacted. And I would have felt dread – dread about having to make the announcement that we had failed. I knew from our early open-label trials that our treatment did show benefit to patients, but if we had chosen the wrong end points our trial would have been considered a failure. We'd be back to the drawing board, and many more people would lose their sight while we tried to fix things.

As a research scientist, I know all too well what failure feels like, and in comparison to the many failures I experienced in the past in my lab this failure would have been orders of magnitude greater.

But in fact I felt incredibly lucky and grateful. I was profoundly grateful to Andrew Strong, who had led the clinical research on

AMD at QLT, and our principal investigator, Neil Bressler, who was at Johns Hopkins. He and Andrew had put together the clinical protocol for the phase III trial and they had done a brilliant job. They had sized the trials appropriately to give us highly significant results, and chosen the right end point. There were still several places for us to fail to capitalize on what we'd accomplished. But we had made it over what most people in the industry consider the toughest hurdle.

We still had several steps to take before announcing our results publicly, and we had to make sure to get each step right. We were scheduled to meet with Ciba and the data safety monitoring committee in Atlanta, where Ciba Vision's US head office was located.

I remember that our son Ben and his family came down from Roberts Creek to Vancouver to spend Christmas with us before we left, and that we had a good time. What I don't remember is how I managed to throw a Christmas dinner and deal with presents for everyone. Ben had three children by then, two girls and a boy, Alex, Graeme, and Emma, aged nine, five, and three respectively. So I must have made time to shop for suitable gifts.

Then we were off to Atlanta. We checked our luggage, something we rarely did – but we had gifts for Jennifer and Paul, whom we would see in New York, where we were heading after Atlanta. Unfortunately our luggage failed to arrive in Atlanta. And no one seemed able to trace where it had gone astray. We didn't get it back until mid-January.

We had lengthy meetings with the data safety monitoring committee, discussing the data and the methods of analysis that we used. The committee had run its own statistics on our data as well, using somewhat different procedures. There were no serious disagreements on interpretation of data.

The atmosphere during the two days we were in Atlanta was tense. I had expected it to be euphoric and congratulatory but people were uncertain, almost as though they were looking for problems. I wondered about this. Our biggest fears had been allayed. We would almost certainly be able to gain FDA approval with these results. We had achieved something together that was going to make life so much better for thousands of people by preventing them from going blind. And yet most of those present seemed nervous, not ready to celebrate. We were almost too close to the data. I suppose everyone was nervous about how the world would receive the news.

We wordsmithed the press release that would go out announcing our topline data. I have seen biotech companies miss great opportunities to increase the value of their companies by mishandling press releases after material events. The wording in a press release can sometimes create an unfavorable impression with investors and analysts, particularly if there is uncertainty implied in the narrative. We had brought in an additional public relations firm to assist us with the release, to make sure the information in it was accurate and left no room for ambiguity, but also provided the data in a positive way. There are always traders out there, short sellers, who make money when a company has bad results. These traders bet on failure and then make money when the stock drops. They will look for any negative nuance they can find in a press release. I'm sure we had our share of short sellers.

I was fascinated by the actual data that we had generated. Some of the findings were unexpected and surprised our investigators. We found that a number of the placebo-treated patients had maintained good visual acuity over the year of treatment. Every retinal specialist we had consulted had stated with certainty that once patients were afflicted with wet AMD, their loss of visual acuity was relentless and fast. Apparently, that was not the case.

Also, we had stratified by age and eye color and found that neither of these factors influenced treatment outcome. We had thought that dark pigment in the iris might block some of the light needed to activate the drug. One other surprising outcome was a gender difference in patients on placebo. Women in the placebo group lost significantly less vision over twelve months than men. In the treatment arm there was no gender difference. We were not able to explain that difference. Are women inherently more optimistic than men? Do they try harder when their eyes are tested?

There was a wealth of information in these trial results that we would mine and analyze in the years to come. I'm a bit of a data nut. I could spend hours playing with data sets. I found a kindred spirit in Andrew. Such "data dredging" would yield a wealth of findings that we could use to enhance our treatments.

We left for New York on December 29, still missing our luggage. The airline had determined that it had been loaded onto another plane, but they had not yet found out where it was. But it was fun the next day shopping at Saks Fifth Avenue for clothing at rock-bottom prices in the post-Christmas sales.

We put out an announcement saying we would be releasing results from our trial on the morning of the January 2. There would be an accompanying press conference by phone. On that day we assembled at the Ciba/Novartis building on Fifth Avenue, about a dozen of us in total. Neil Bressler, our principal investigator, took over the conference by presenting our topline results, providing some details. It was a professional presentation. There were a large number of attendees and Neil was peppered with questions, which he answered with clarity and the right amount of detail. In conference calls of this nature, all attendees' names and affiliations are recorded and all questioners are identified before they are permitted to ask a question, so we had a pretty good idea of who was listening in. There were over fifty people on the call. It was exciting to realize that we had received the attention of many of the major investment banks in the United States.

We still weren't sure how the market would react to the news. For Ciba, it didn't really matter; they were part of an enormous pharmaceutical company whose stock price was unlikely to move much on news of our data. But for QLT, the reaction could make or break us.

That day, our stock started to climb, and the volume of shares traded went up from tens of thousands per day to hundreds of thousands per day. Before we announced, our shares had climbed a bit to the low twenties in US dollars. On the day of our announcement, the shares climbed another ten dollars US. The value or "market cap" of a publicly traded company is calculated simply by multiplying the stock price by the number of shares issued in total by the company, so a rise such as we saw added millions to the value of the company.

We'd come through for all the people who had believed in us and invested in our company. It was a great feeling.

The atmosphere at QLT when we got back was something like bliss. People couldn't stop smiling. Some weeks earlier, all the employees had been given yearly grants of stock options, so everyone gleefully watched our share price rise steadily over the next few weeks. We were getting to dizzying heights, for us, with shares hitting forty dollars US. Employees started exercising their options. And for the next little while I'd never know what I was going to hear when I got in the elevator. One of our executive assistants told me she'd bought a condo, something she never thought she'd be able to do in Vancouver, one of the priciest real-estate markets in the world. Another young man in our clinical department, whom we knew was HIV positive, told me he was taking his parents on a trip to Paris, something they'd always dreamed of. People were taking their kids to Disneyland or getting their kids the things they'd really wanted for Christmas that had been considered unaffordable. Someone else bought a Porsche. When I reacted negatively to this news, the young man who told me got embarrassed and tried to make excuses for his excess. Everyone felt part of what we had been able to do and it was a very good feeling.

A number of significant investment analysts from US banks started coverage of the company. Their reports served to push the stock even higher. We started getting invitations for meetings with substantial fund managers. It was pressure from our US analysts that convinced us that we should issue more shares through another public offering, this time mainly in the United States and led by a US bank. Our stock, according to some of the analysts, wasn't liquid enough. Big investors wanted to be able to buy large numbers of shares, but if they started buying the price of the stock would rise too fast. We had been a little-known Canadian company before this time, and one of the US analysts now following the company remarked that we were the only biotech company he'd known that reached a market value of over a billion dollars without having a single US analyst following it. That had all changed now.

We debated internally whether to do another offering. We didn't actually need the money as far as we could judge. We had enough to see us through the filing with the FDA. But pressure continued, so we agreed. It would be a new experience, being taken around by US analysts. We'd done extremely well by our Canadian bankers and felt a bit traitorous picking another bank to lead the offering. Some of our Canadian shareholders were very angry about our decision to issue more shares. They were delighted at the way the stock had performed and worried that a dilutive event like a big share offering would serve to weaken the value of the stock they held. As it turned out, the offering did nothing to hurt the share price, which continued to rise.

This road show in the United States was the most rigorous and exhausting marathon I have ever been through. We crisscrossed the continent using private jets because our schedules were so tight we didn't have time to use commercial airlines. We flew to Europe for a whirlwind tour of London, Paris, Lausanne, and Copenhagen. At the end of the three-week duration, our share price had held firm at between forty and forty-five dollars US and we had a high demand for the stock. We raised a hundred and sixty million dollars.

The first corporate jet we rented took us from Vancouver to Chicago and from there to New York after meetings in Chicago. We boarded from a building separate from the main airport and were spared the lineups and checkpoints. The jet was small, with seats for the pilot and co-pilot and six more for passengers. I was surprised that there was no toilet available, but when we asked, the pilot smiled and pointed out the seat Ken (our CFO) was sitting on. If one removed the cushions from his seat, there was a small toilet basin. Obviously these jets were built for single-gender parties. I quickly got off the jet and used the facilities in the building. With the advent of more and more women in banking and corporate management, these jets must have been replaced by now.

Not all corporate jets were so spartan, fortunately. We flew from San Francisco to New York later in the trip and that jet had a miniscule private toilet, although during some severe bumpiness the young woman who was our hostess threw up messily in it and rendered it almost unusable.

Our trip was wildly successful and our stock continued to go up, even with the dilution that so many of our Canadian shareholders had feared. In the three weeks of endless meetings, travel, and long hours, we were exhausted but happy. We felt like stars or real VIPs, clearly an illusion of the overtired brain as well as an effect of all the lavish praise we had received. Our stock had reached about seventy dollars US.

On the last day of the road show, I flew directly from where we were in Texas to Fort Lauderdale for a big ophthalmology meeting where Neil was going to present our data at a major symposium. The large room was packed, and Neil produced his usual polished presentation of our results. It was very exciting to see the interest and enthusiasm of the retinal specialists attending.

Back at QLT, the onerous task of putting together the FDA filing for approval of Visudyne was under way. We had good data and had shown a significant patient benefit, but that did not provide an excuse to put in a shoddy, incomplete filing. The spirit around the company was high-energy and no one was shying away from the hard work. Alex Mancini, as head of regulatory, was organizing the way in which the filing was put together. Andrew was leading the clinical part of the filing, so the animus between Alex and Mohammad was not an issue. Mohammad was occupied with the Barrett's esophagus trial.

Nobody in any of the divisions – clinical, manufacturing, devices, preclinical, or toxicology – wanted to be the group that was last finishing their documentation, so everyone worked hard and competitively.

The people in manufacturing were confident they would have their documentation done ahead of other divisions. Then someone did that extra experiment that should not have been done. Once a process is "frozen," there should be no tampering with it. The issue involved one of the standard assays for the drug itself. Visudyne is a mixture of two almost identical molecules called regioisomers. They had the same chemical makeup and the same activity, but one of the side chains of the molecule was oriented slightly differently from the other during its synthesis. This is not unusual in organic molecules. The two isomers are separable using standard chromatography procedures and show up on a graph as two separate peaks of the same size. They had been shown to have identical biological properties. One of our scientists was looking at the two peaks and decided that if he changed the buffers slightly he might be able to produce a cleaner profile and sharper peaks. He started playing around and, to his surprise, when he ran the Visudyne preparation through his new buffering system, not only did he sharpen up the two peaks, but also two small blips on the graph showed up as contaminants from underneath one of the two major peaks containing the active molecules. He realized these constituted impurities in the product that had not been identified before. He would probably have liked to disregard what he had seen. But in an ethical drug company following good manufacturing practices, one can't "forget" such a finding. The two impurities that constituted a tiny percentage of the product (far less than 1.0 per cent) had to be documented and characterized. The chemist who had done this extra piece of work did not win any popularity contests that day.

Not only did the company have to characterize these two components chemically, we had to show that they had been present in all previous preparations of the product. If they were "new" impurities that had been absent in earlier batches of Visudyne, that could mean that all our toxicology work would have to be repeated. All sorts of unpleasant scenarios surfaced. Our filing could be seriously delayed.

The manufacturing group went to work, checking all previous batches. Fortunately, the two impurities showed up in all the earlier batches and turned out to be molecules similar to our active components that were formed during the manufacturing process and had already been identified by the analytical chemists. But the manufacturing group had been delayed because of this small unnecessary act of "fine-tuning."

That summer, Ciba Vision and QLT were awarded the Helen Keller award for the work we had done on Visudyne for fighting blindness. It was a proud moment when Gusti and I received the award on behalf of both companies.

Our company expanded dramatically during the six months it took to complete the documentation for filing for approval of Visudyne. Linda Lupini and her staff were constantly interviewing and hiring. People who have had experience in drug development were pretty rare in a place like Vancouver, which, until QLT was formed, had no biotech or pharma R & D base. All senior staff had to be imported from either the United States or Europe, and occasionally from eastern Canada, which unlike western Canada had a few biotech companies and several large pharma company branch plants. But we were able to hire local people with experience in manufacturing, quality control, and quality assurance as well as science and clinical research. Somehow, we succeeded in putting together a great team of people.

By midsummer, Alex and Andrew, with the assistance of dozens of QLT employees, had put together the documentation for filing. All the information was stored in four-inch binders. Three hundred of them made up the filing. It had taken a couple of dozen people six months of hard work to assemble the documents.

We celebrated as a company the day we finally sent the filing off to the FDA. The documents filled up a large post-office delivery van.

While we had been preparing the documentation for the filing, we were sending all of the processed information to our Ciba partners. Ciba Vision regulatory staff were now working with the Novartis regulatory people at their head office in Basel preparing the European filing. The EU had established a process by which filing for a new drug approval in Europe no longer had to be done on a country-by-country basis, as it had been in the past. Even so, approvals were in the jurisdiction of individual countries, so it was possible to be approved in one country but not in another.

Then there were issues of pricing and reimbursement for the drug, which did have to be negotiated on a country-by-country basis. So while the lion's share of the work had been done, I learned that there were many more steps to be negotiated before we would have a successful launch of the product. But all the remaining issues to be resolved were in the hands of our partners, because they pertained to marketing and Ciba was in charge of that.

One thing played in our favor, and that was that the vast majority of patients receiving Visudyne in the US would be over sixty-five and therefore eligible for government-covered Medicare, so negotiation for reimbursement would be only with one main third-party payer. Reimbursement for new treatments in Canada has to be negotiated on a province-by-province basis, and that would prove to be difficult.

We had been an R & D company, but now we were one that had revenues. Most of us at QLT had no knowledge of what that was going to be like. But where we were at that moment felt pretty good.

When I thought about what we had accomplished I sometimes felt overwhelmed. We had been incredibly lucky. There are so many ways to fail in something as complex as drug development. Somehow, we had dodged a fusillade of bullets. And soon thousands of people would benefit from our treatment worldwide. So many people at QLT and Ciba Vision had made significant contributions to the development of Visudyne. I felt intense gratitude to all of them.

CHAPTER TWENTY-THREE

More Than a One-Trick Pony

We had very little opportunity to rest on our laurels. Our investors were happy with what we had accomplished. At the same time, they became demanding. We started getting comments and questions such as, What about your pipeline? How are you going to fill it? I hope you're more than a one trick pony. How are you going to increase shareholder value? How soon is your filing going in?

I couldn't help feeling resentful about these questions. We had given our all collectively and had performed well. But we had to pay attention. As senior employees of a publicly traded company with big influential shareholders, we had a responsibility to listen to them. A lot of individuals had also invested in us, and we had a duty to them, too. But they were not so demanding. At every annual general meeting I was always gratified to have people come up and tell me how interested they were in our company and how pleased they were to be shareholders. But other shareholders are more difficult to please. Growth is considered essential for successful companies, and that meant full pipelines coupled with a continued rise in stock price.

It was true; our pipeline of near-term products did not look robust. Our staff at QLT had focused almost all their efforts on getting Visudyne through clinical trials. Everything else had taken a back seat. We didn't have the kind of exciting prospects that the market expected. Some of our researchers were working on a next-generation photosensitizer as well as putting together information so that we could decide where to take the verteporfin (Visudyne) molecule beyond ophthalmology. We did have Photofrin in clinical trials for Barrett's esophagus, but our investors were not interested in that. While Barrett's esophagus was potentially an important indication for PDT, Photofrin sales were never expected to affect our stock price. Photofrin would never be a big product. Now we were cursed with having a potential blockbuster drug on our hands, and nothing but another potential blockbuster in the pipeline would satisfy the market.

In order to satisfy investors, we should have had products already in phase I and II clinical trials and other potential products in preclinical development. That meant buying another company or in-licensing later stage products. Fortunately, we had a lot of money in the bank and stood to have more once Visudyne sales began, so we could look at other products.

Our shareholders wanted us to use our financial position wisely. The company was going to have to change. I could sense the anticipation of our shareholders and employees. But making the right choices on how to change was a daunting responsibility and full of risk. I wasn't at all sure we had the expertise to make the right decisions as these changes took place.

We now had over four hundred employees, and we were not as nimble as we had been when we were smaller. And adjusting to change was critical if we were to succeed.

One of the major changes that took place that year was that Alex Mancini, our vice president of regulatory affairs, left the company shortly after we had filed with the FDA for approval of Visudyne. Alex had been one of the first people we hired who understood and taught us how to deal with the FDA and the complexities of regulatory issues. In our sometimes-fractious relationship with American Cyanamid, she had been a voice of sanity. She was the key person who had managed to get Photofrin approved by the FDA. Her seemingly tireless efforts and small stature had earned her the title of "Energizer bunny" in the company. We owed her a great deal and we were losing someone who had a deep understanding of PDT. We were fortunate that Alex's counterpart at Ciba Vision, Larry Mandt, was located in Atlanta and was familiar with and able to handle the Visudyne file. While Larry didn't have Alex's PDT experience, he was a valuable asset in that he was very familiar with the ophthalmic division of the FDA.

I realized that QLT needed a chief scientific officer. I'd been filling that responsibility since I became CEO, but if we were going to be doing serious due diligence on other companies and products, we needed a CSO. John North, our Beaufour liaison, had impressed me with his critical thinking capability and his eager support of the PDT field. He also had a deep and analytical understanding of drug development. He had let me know he was thinking about leaving Beaufour and was looking for other opportunities. When I told him I was looking for a CSO, he jumped at the opportunity.

In retrospect, I realize how naïve most of us at QLT were at this time. We were certainly excited by what we had achieved, and filled with self-confidence. We had beaten the odds once and we could do it again. We would find a way to grow the company. Lee Anne and Ed put together ad hoc committees to do due diligence on potential takeover possibilities or in-licensing opportunities from other biotech companies.

Another issue loomed in our ophthalmology franchise. It had been known for some time, from research in ocular models of macular degeneration, that a human growth factor called vascular endothelial growth factor (VEGF) likely played a role in causing the growth of the leaky blood vessels characteristic of wet AMD. VEGF is a promoter of new blood vessels and is important in tissue repair. But in some environments it can have harmful effects. There were a number of VEGF inhibitors already on the market or being tested clinically.

Genentech's Avastin was such an inhibitor. It was a monoclonal antibody to VEGF and had been approved for the treatment of metastatic gastrointestinal cancer. VEGF plays a key role in stimulating the growth and development of blood vessels, and metastatic cancers require new blood vessels to establish themselves in the sites they have migrated to. Inhibition of VEGF had been shown to slow the growth of metastatic cancers and prolong life in some patients.

We thought about approaching Genentech about a possible collaboration. Avastin was still on patent and would not be available to us without an agreement with Genentech. We did speak with Genentech casually at one point and got the distinct impression that they were not interested in talking to us. Later, we were to discover Genentech didn't want to collaborate with us because they wanted to compete. Genentech was already planning to become our major competitor at that time.

We knew that VEGF was produced after PDT as a response to the photodynamic effect that closed the leaky blood vessels. It was probably the reason so many patients on our treatment had recurrences. It seemed obvious that a combination therapy with Visudyne and a VEGF inhibitor (also called an antiangiogenic) might reduce the number of recurrences and thus the number of treatments a patient would need. That would mean less revenue to the companies; but we at QLT were interested in providing the best possible treatment for patients.

We discussed the principles of planning a combination therapy trial with our Ciba partners. Luzi, I could tell, was dragging his feet, but Gusti was on board. It turned out that Novartis, the parent company of Ciba Vision, had a VEGF inhibitor that might be available to us. It was in early-stage clinical testing for cancer but could be made available to Ciba Vision at no cost. Luzi thought we should wait until the Novartis trial was over, a few months from then, and decide at that time on how to proceed with a combination trial using it.

We had restrictions as to what we could do about expanding the Visudyne franchise in ophthalmology. Our agreement with Ciba Vision mandated that both companies had to agree on additional studies. So, although most of us at QLT would have preferred to in-license a more thoroughly tested anti-VEGF reagent and get going immediately on testing the combination therapy, we went along with Ciba Vision, even though we had heard unfavorable things about the Novartis drug and didn't have high expectations for it. We thought we could afford to wait a few months before choosing an anti-VEGF drug for a combination study.

The Novartis drug had been developed with the aim of treating cancer, and we had heard the side effects were fairly severe. Such a side-effect profile might be acceptable for people suffering from cancer, but not in otherwise healthy people with AMD. Our fears were borne out when the trial results for the Novartis drug became known. Side effects included explosive diarrhea and projectile vomiting. So we were back at square one looking for a VEGF inhibitor.

Meanwhile, the due diligence teams within QLT were identifying other possible in-licensing opportunities that included VEGF inhibitors. We tasked our teams with different missions. One was to find VEGF inhibitors, another to look at merger possibilities, and another to look at in-licensing opportunities in the area of cancer or immunotherapies. Each team had people from marketing, clinical, regulatory, preclinical, and business development, so that all aspects of a possible acquisition could be examined.

The dot com bubble hadn't yet burst in 1999 and many companies, biotech included, were grossly overvalued, including QLT. Ever since we'd filed for Visudyne approval in mid-year, the book value of the company had been in the billions of dollars. But many other biotech companies were also valued in this range, so trying to justify acquisition of another company would be costly.

After she left QLT, Alex Mancini joined Inex, another Vancouver-based biotech company, which Jim Miller had started after he left QLT. Alex took charge of their first product, called Marqibo, which was a liposomal formulation of vincristine, a cancer drug that had been on the market for some time. Inex was running a phase III clinical trial in non-Hodgkin's lymphoma. After having put such enormous and successful effort into getting Photofrin approved in many jurisdictions, Alex would now turn to another difficult challenge in Marqibo and try to work her magic on it to get an FDA approval. We did some research on this product because it was possible we could partner with Inex. The idea of partnering or merging with a local company was attractive to me. Strengthening the local biotech community has always been an interest of mine. However, when we looked in detail at Marqibo, we were unimpressed by the quality of the product. Unfortunately for Inex, when the time came to file for approval of Marqibo, even Alex's wizardry was not sufficient to capture the imagination of the FDA on this product and the filing was refused.

We looked at a number of other local biotech companies, as well as Seattle-based ones, reasoning that another company in the same time zone or geographical location would be easier to manage. By far our favorite pick was a company call Anormed, located just outside Vancouver in the Fraser Valley. I knew the company well as I had sat on their board for some years since they moved from the United

States to Canada. I knew and respected the management of Anormed and I knew their science was very solid. Their CEO, Mike Abrams, was a scientist like myself and we had a lot in common. Their corporate culture was similar to ours. They had been through some tough times, similar to ours, but were back on track. They had a product that was about to enter phase III trials, a product that would help patients recover after receiving bone marrow transplants. They had an interesting pipeline. The two companies, QLT and Anormed, had complementarity. They had an excellent medicinal chemistry group, whereas we were strong in biological models and had first-rate animal facilities.

I was sixty-five that year and was looking to step back from my leadership role once Visudyne was on the market. Mike Abrams, the CEO of Anormed, was younger than I, a seasoned CEO, and could be my successor. I thought investors would approve of such a union.

Ken and I arranged to meet with Mike and Bill Adams, Anormed's chief financial officer, and broached the subject of a merger with them. Mike and Bill had known what was coming and had clearly discussed and thought about the possibility. They politely refused. Mike told us he wanted to see if he could make it on his own. If they had merged with us, we could and would have financed his phase III trials and we could support his pipeline. Life would have been much easier for the Anormed employees, and Mike and Bill knew that.

But I understood them. They were still enjoying the excitement of bringing a product through testing and could taste the moment of success. They were excited about raising money on their own. They weren't ready to take an easy way out. There was no ill will generated by their refusal.

So we didn't succeed in what would have been a successful merger. Many years later, after Mike had gone through a hostile takeover orchestrated by activist shareholders, he admitted to me that he regarded his refusal to merge with QLT as the worst mistake he'd made in his corporate career.

For the remainder of 1999 we looked at a number of possible opportunities, and ended up having doubts about the science behind the technology. We wondered if we were being too critical. In retrospect, I realized we were correct in our evaluations, since in many instances the products we rejected failed in later clinical testing.

About this time, our management group decided we should go from matrix management to project management, something that most large pharmaceutical companies used. When QLT was growing, people we hired were put into individual departments like regulatory, clinical, research, manufacturing, and so on. Their department heads then assigned them to various ongoing projects, but they still reported to the department head. With project management, when a new project was brought into the company or a research project moved from research into development, a project manager was assigned to take charge of that project. Project managers then put together a project team involving individuals from each department to undertake the various tasks required to move the product forward. They negotiated with department heads for the time of those chosen individuals from the departments. Project managers have to be conversant with all aspects of product development and have to play the part of coach/cheerleader for people within the project. Good project managers have to have a wide breadth of knowledge and be skilled diplomats to get the most out of their teams.

While QLT had performed most of the work in Visudyne development, now the workload for the drug became the responsibility of Ciba Vision and their marketing division. At QLT, Lee Anne had hired Bob Butchovsky, who had been in middle management at a California-based ophthalmic company, to be QLT's main marketing person for Visudyne. He worked in concert with the Ciba marketing team and coordinated any participation QLT might offer. Preparing to launch a new product, especially one that is unique, involves considerable expense and preparation. Sales reps had to be trained in the new technology, and seminars, to which ophthalmologists are invited, had to be organized. These are usually held in luxury resorts where doctors are treated to lavish meals and cocktail parties. I began to understand why marketing budgets were always so high.

Towards the end of December 1999 we received our first approval for Visudyne. The approval came from Switzerland. We had started our development program with Ciba Vision in 1995. By biotech standards, we had taken Visudyne through its development path at lightning speed. I believe this spoke to the unmet need of AMD patients as well as the efficacy of the treatment and the careful planning of our clinical team.

Photofrin continued to be problematic. Sanofi was not happy about the difficulties they were having marketing the drug for lung and esophageal cancer. And at Beaufour Ipsen, Dr. Beaufour was very elderly and had withdrawn from all involvement with the company, and J.P. lost no time terminating our research and development agreement. We really didn't mind so much. He had been a very difficult person to deal with. I strongly suspected that our partnership with Beaufour would not be long-term and that we'd soon have Photofrin back.

Our daughter Jennifer and her husband, Paul Schwalb, had decided to come to Vancouver for the holidays. Ed and I both liked Paul enormously and felt they would be happy together. They were aligned in their political views and complementary in personality. But we realized it was now unlikely Jennifer would be returning to Canada in the near future, and that saddened us.

Jennifer worked for Brooklyn Legal Services in their housing division. She had become involved with homelessness when she was an undergraduate at New York University, so this employment was a logical extension of her interest in helping the struggling poor.

Jennifer always had a passionate love for our summer property on Sonora. The millennial new year was just around the corner and we decided to celebrate it off the grid. Weather is very uncertain at that time of year, especially around Sonora Island, which is located at the mouth of Bute Inlet. Winds from the arctic can come whistling down the inlet and cause dramatic drops in temperature and brutal gales that down trees. Alternately, mild weather could prevail, which usually meant being drenched in rain.

That year, temperate weather blessed us with mild rain. Several other families had decided to see the new year in at Sonora too, so we had a memorable New Year's Eve celebration.

The things that end up in cabins in remote locations are like a time capsule. Things that get brought never get taken back and never get discarded. We had many clothing relics from the seventies. Ghastly muumuus with psychedelic designs, harem pants, and lots of paisley shirts and ties were all stored in musty drawers. We dressed up in whatever bizarre get-ups we could put together and convened around a sumptuous potluck, wondering if, at the stroke of midnight, planes would drop out of the sky and the world come unstuck. The predicted millennial calamities couldn't affect us at Sonora. We were disconnected from any grid. We spent a few days on the island, sitting around our fire and playing bridge. It was a delightful respite from our regular lives.

Then we returned to the real world. Jen and Paul left, and we were back at QLT, excited about what the year might bring us. Our shareholders were all focused now on what would happen at the FDA and how successful we would be at getting Visudyne to market after its approval in the United States. Although we thought we were fully prepared for the launch of the product, I was aware that many things could go wrong.

CHAPTER TWENTY-FOUR

Launching the Product and Hoping for No Unpleasant Surprises

In mid-1999 we submitted our results from our AMD trials to the ophthalmic division of the FDA in order to obtain their approval for us to market Visudyne in the United States. We knew the FDA would need six to eight months to carry out their review of the information. We had submitted information on a new drug and a new method of treating macular degeneration. When the FDA reviews applications with new drugs, they invariably require that the sponsor (the company) appear before an advisory committee to answer questions.

This was our situation facing the new millennium. We embraced 2000 with excitement, anticipating the FDA meeting and the subsequent approval of Visudyne. Once we had that, we could start marketing Visudyne in the United States. Ciba Vision marketing was geared up to launch the product in the United States as soon as we received that clearance.

In order to prepare for a successful product launch, we explored ways to get lasers placed in physicians' offices ahead of the FDA approval. In collaboration with the two companies we had enlisted to produce the lasers, we went about placing them in a majority of retinal specialists' offices. Retinal specialists are fairly rare, and many of them have group practices. Retinal practitioners treat the back of the eye for diseases like diabetic retinopathy and tumors, so they are familiar with lasers and ophthalmic surgery. The lasers they used prior to our PDT lasers were mainly thermal, which destroyed diseased parts of the retina by using heat. So our PDT lasers were something they had no difficulty dealing with and they were not deterred by the cost of placing another laser in their operating facilities. We calculated that four hundred of the devices could provide laser access for the retinal community in the United States. Ciba and the laser companies marketed the lasers to the retinal practices ahead of our obtaining FDA approval, so many physicians were ready to start treating before we had approval. We had trained the sales staff in the PDT procedure and provided short courses for physicians, so a lot of groundwork was finished before we got the approval.

The ophthalmology division of the FDA is small in comparison to the oncology division. It was run at the time by a man called Wiley Chambers, an ophthalmologist himself, who had a reputation for being unpredictable, tough, and independent. Larry Mandt, who had replaced Alex as our chief of regulatory affairs, was familiar with him, having been in front of Chambers a number of times when he was at Ciba Vision. He said he found Chambers enigmatic but didn't feel he had a bad relationship with him.

We heard from Chambers in February and were given the date for our advisory committee meeting and the names of the people they had appointed to the committee. Many of them were known to us. We were familiar with the retinal community and had recruited many of the key retinal opinion leaders to participate in our trials. So it was not surprising that we knew several of the people selected for the advisory committee.

We brought Neil Bressler in as our main spokesperson and made preparations for our presentation. We were sure we would get approval. After all, we had carried out the trials in accordance with FDA standards. And we had achieved positive outcomes with every end point agreed to with the FDA. But even that knowledge didn't prevent us from becoming very nervous in the lead up to the meeting. This hearing would be well attended by investors and analysts and we didn't want to be embarrassed.

Advisory committee meetings are usually held in one of the hotels in the environs of the FDA in Bethesda – Hiltons and Holiday

Inns abound in the neighborhood. Our Visudyne meeting was held in one of the large hotel ballrooms to provide accommodation for the expected audience. And most of the seats were occupied. All the analysts who followed our company as well as those covering other companies that were our competitors were in attendance.

There were no unpleasant surprises. Neil was professional and smooth in his presentation. He had such enormous respect from the retinal specialists in North America and elsewhere that we had come to count on him. Wiley Chambers then presented the FDA analysis of our data. I could see that he had taken a lot of trouble with our documents and had done a number of independent statistical analyses of the data. I realized that our product was radically different from the usual ophthalmic products that came across Wiley Chambers's desk, which consisted mainly of products for cleaning contact lenses or eye drop formulations of already approved drugs. Not only was our product a new chemical entity, it was also formulated for intravenous administration and was a product for treating the back of the eye or retina. All these facts were firsts for the ophthalmology division of FDA.

So it was not surprising that Chambers took time to go over our data thoroughly. But in sum, there were no real substantive differences between his findings and ours. The advisory committee also presented no real difficulties for us in the questions they posed. We received an unopposed recommendation for approval.

Our stock price rose again on the positive news.

After the advisory committee, all we had to do was wait for the FDA to send us an approval letter. I'm not sure what takes the time once the advisory committee is over, especially when there have been no obvious problems. But it took about three weeks before we received our approval. The approval letter arrived late afternoon via fax. The press release had already been written and it was sent out immediately.

Ciba Vision started selling Visudyne the next day. Neil Bressler received the first commercial vial of Visudyne, which was then enshrined at Johns Hopkins Hospital. He also treated the first patient with the approved drug. Sales that first day topped a million dollars. That meant about a thousand vials of the drug were sold. The vast majority of those patients were covered by US Medicare, so the charge to a patient who did not carry any additional insurance would not be high.

The Ciba sales force was very professional. The time and money they had spent before we got approval was paying off now. Ophthalmologists had been trained and were primed to start treating patients. AMD is a serious condition and there was a pent-up need for therapy. Sales of the product during the first three months after approval in the United States were brisk, and when the time came for us to provide information on our financial status for the second quarter of 2000 we showed a profit for the first time in our twenty-year history.

At that time there were very few profitable biotech companies. We were certainly among the first in Canada, after Biochem Pharma in Montreal. And we were about to learn about the difficulties of being a profitable biotech company.

I had naïvely thought that once we started making money we would be at liberty to invest heavily in our own research and development pipeline. Then Ken, our chief financial officer, gave me a quick lesson in the expected governance of publicly traded profitable companies.

I must admit, I'd never paid a lot of attention to our quarterly financial reporting. Ken was so proficient and dealt so professionally with investor questions about our finances that I'd simply left these things up to him, paying attention mainly to his responses when he'd been asked questions by investors. Balance sheets have never thrilled me. I have sat on many boards over the years and have even been coerced occasionally into sitting on audit committees, and even now I still find myself glazing over at meetings when the minutiae of finances are presented and discussed. I can never get over how excited some of my fellow board members get when the balance sheets come up on the agenda. I was never irresponsible about spending myself when I was operating at the university from grants. I always paid attention to how much money we had in the bank and whether we could support the activities of the company, but the intervening numbers above the bottom line did not hold my interest. I have always been driven not by money but by scientific achievement.

I knew that we reported our financial status as earnings per share (EPS) every quarter. Up until 2000, our EPS was always registered as our loss per share, which was understandable. We were spending money and not earning any. But now we recorded positive EPS as earnings. The stock market pays enormous attention to EPS, which is a measure of how well the company is managing its profits and how well its products are doing. I learned that there were unwritten rules about quarterly expenditure. The investment community frowned on companies that expended much more than 25 per cent of their earnings on research and development. So we had to keep our R & D expenses under control. This was the most disappointing discovery I made about profitability. And I started recognizing the importance of some of the lines and numbers on balance sheets. For the first time I discovered the drawbacks of being a publicly held profitable company and having influential shareholders.

I also discovered that once a company becomes profitable it is judged differently by investors. They watch the EPS every quarter and use a simple metric for judgment. If the EPS goes up quarterly, the company is doing well and justifies investment in it. If the EPS goes down, they believe the company is not being properly managed and in some extreme circumstances shareholder action can ensue. Increased R & D expenditure lowers EPS. This has always struck me as a very questionable way to value biotech companies, which are fueled by the original research that they carry out. I never understood how investors complained on the one hand about our pipeline being weak and then complained on the other about the amount of money we spent on R & D.

We had a lot of money in the bank, but many investors only care about what percentage of your earnings you were spending on ongoing R & D and other running expenses and whether your profits increased quarter over quarter. You would be punished if you exceeded 25 per cent of earnings on research. What you could spend that money on was making acquisitions or in-licensing. That was clearly what we had to do to satisfy the voracious appetite of the shareholders.

At the same time, we still had the troublesome problem of Photofrin on our plate. Our marketing partners were unhappy with the sluggish performance of the product. Lee Anne approached me at this time with the suggestion that we build a small sales force and market the drug ourselves. She had not discussed this possibility with anyone else in the company and she made it clear that if

I didn't support her suggestion she'd take it no further. We talked at some length about her idea. I was filled with doubts. I was still very uncomfortable around sales and marketing. It was still alien to me.

I had been astonished by how much money Ciba Vision spent leading up to the Visudyne launch, and the high number of people involved in the sales force. And I know that by most standards Ciba was considered frugal. Hiring a sales force would drastically change the character of QLT. Lee Anne assured me she would handle the marketing of Photofrin very differently. After all, the product was already on the market. She would hire a small dedicated sales force that would be responsible for sales in the United States and Canada. Beaufour was still marketing the drug in Europe, even though our R & D agreement had lapsed.

I saw how much Lee Anne wanted to do this. After all, she was a marketer and she had worked for a company with no active marketing sales force for years now. But I thought about all the new challenges we were facing with in-licensing and acquisitions and couldn't see the wisdom of embarking on yet another unfamiliar challenge. I told her so but said I had no objection to her taking her idea to the management committee and seeing how it went with the others. I would abstain from the discussion and go with a majority decision. She said she wasn't going to do that. We should just drop the subject.

I realized she didn't want to put her ideas to the management committee unless she already had my support. Lee Anne didn't like to lose, and she knew she could win with my support and was uncertain of the outcome without it. I saw how disappointed she was and felt I'd let her down. I would never know for certain if my opinion was the right one at that moment, but subsequent events led me to believe I'd probably made the right decision.

We decided we should try to sell Photofrin to an appropriate company whose business covered clinical endoscopy practices. We let it be known that we were interested in doing so and were approached by a Montreal-based company called Axcan. It was a company with no R & D pipeline and was focused on marketing gastrointestinal drugs and devices. Most of their products had been in-licensed. They seemed an ideal match for Photofrin. We were now in the middle of the Barrett's esophagus Photofrin trial and Axcan was keen to include this disorder as part of the Photofrin portfolio. We came to an agreement with them that we would complete the Barrett's trial for them and file for approval in the United States and Canada as part of the deal to sell Photofrin to them.

Saying good-bye to Photofrin was bittersweet. The product itself had given us lots of grief. But we had cut our teeth on Photofrin and learned how to go about drug development with the American Cyanamid partnership. We would never have been able to develop Visudyne as successfully as we had without the learning experience of Photofrin.

Tom Dougherty, the creator of Photofrin, was very upset by our action. He had felt comfortable with our management of the product, and Axcan was an unknown. Tom and I had a warm relationship. I hated seeing his disappointment and was saddened. When I talked to him later, he understood our position, and we have remained good friends over the years.

Another thing I learned about being a profitable company was the ritual of sales projections. I had heard market reports and sales projections for many years but had never equated them to anything I was involved in. Now those projections became very real and significant. Public companies are obliged to make projections every quarter about what their total sales and EPS are likely to be quarter over quarter. Large companies with many products on the market have ways of buffering against unpleasant surprises from one product or another. But we literally had nowhere to hide. Our quarterly financial returns were driven solely by Visudyne sales. And we didn't even have control over those. I would have liked to under-promise and over-deliver, but Ciba Vision was a slave to its Swiss masters, who wanted accurate estimates and got points for being as close as they could be to those projections. The marketing people track sales of individual products on a daily basis. Bob Butchofsky, our Visudyne marketing director, knew week by week how sales were going. Reporting of quarterly earnings to the public takes place two to three weeks after the end of each quarter. Our quarterly finances could only be determined after Ciba tallied up the total sales and subtracted the marketing costs, which was a complicated process once they started selling Visudyne in many countries. We were at their mercy, and that was not a good feeling.

During this time, the two PDT companies that had followed us into AMD, PDT Inc. and Pharmacyclics, both reported that their phase III clinical trials had failed to meet their primary endpoints. The PDT Inc. drug had been poorly formulated, took a long time to inject, and did not have the optimal pharmacokinetics to produce the kind of closure of the leaky blood vessels needed for a good clinical outcome. On top of that, the drug caused prolonged skin photosensitivity. The Pharmacyclics drug was water-soluble and did not accumulate adequately in the leaky blood vessel walls. They had used concentrations of drug so high that it had produced negative neurological side effects like finger numbness and tingling. Also, the skin of treated patients took on a bluish tinge like the color of the drug.

I have to admit that I was relieved that we didn't have a competitor nipping at our heels. But I felt sorry for the companies that had failed. I knew both the CEOs and many of their scientists, because we all attended the same international PDT meetings.

The scientists who led the development of both these products were chemists. And neither of them appreciated the importance of a deep knowledge of the physiological and pharmacological properties of their drug. The combination of Anna Richter's and my biological background, combined with David Dolphin's chemical genius, was what had distinguished our product from theirs. We dug as deep as we could to understand every biological property of the molecule. I had had several graduate students do their PhD research on the Visudyne molecule's behavior in the body, funded by university-industry grants from the federal government. We knew exactly what happened when our liposomal drug was injected intravenously, how it broke from the liposomes and distributed to cells. We knew why it accumulated in the walls of new blood vessels. I believe it was our diligence in digging deep into the in vivo characteristics of our molecule and our knowledge of where it distributed in the body that distinguished it from the other photosensitizer molecules in development, and would later lead us to the success that eluded others.

Regardless of the uncertainties about sales projections, 2000 rolled out as a banner year for QLT. When we reported our second-quarter earnings that September, the sales of Visudyne far exceeded Ciba-Vision's projections and those of the market. Our share price that year jumped to $120 Canadian from between $80 and $90, and the volume of shares traded in the United States was in the hundreds of thousands every day. The large investment firms in New York and Boston all started paying attention and our company was picked up by a number of top-flight analysts. We started getting invitations to attend and give presentations at all the major American investment bank meetings, including those of Goldman Sachs, Alex Brown, Oppenheimer, Lehmann Brothers, Citibank, and others.

Being a new face in this world of cynical finance is like being the most popular girl at the prom. Everyone wants to get on your dance card. The meetings are similar in some ways to scientific meetings in that companies present their information in seminar form to a wide audience and provide not just the scientific information but the financial as well. Ken and I did these presentations together, I giving the science and he giving the financial information. Then there were a series of one-on-ones with individual potential investors, who want a private audience. We found ourselves with almost impossible timetables, with the one-on-ones scheduled every thirty minutes all day. It got incredibly boring providing the same information over and over again. I diverted myself by trying to figure out what each person we met with was most interested in and altering my presentation to provide additional information in those areas of interest. I felt sorry for Elayne Wandler, our director of investor relations, who was obliged to sit through these mind-numbing performances, keeping us on time, making sure everything ran smoothly. She was the one who had cultivated the relationships with these investors and she and her staff spent time with them on the phone in the time between meetings.

Investment bankers can make you feel very special and very important when your stock price is flying, as ours was. We were courted with dinners at expensive restaurants, invitations to the theater and sporting events. We were moving in the circles of the masters of the universe and it was a very heady experience.

These masters of the universe are an interesting group. Having dealt mainly with Canadian investment bankers and their analysts up until now, I had formed an overall impression. I'd noticed that in Canada the analysts, as opposed to bankers and fund managers, are quite varied and impossible to generalize about. They were certainly all intelligent, some more than others, and they were serious.

Montreal and Toronto are the main locations for the major money institutions and pension funds in the country. Amongst the bankers, I found the Montreal scene much livelier than Toronto, and in Montreal we were more likely to be taken to better restaurants.

I had trouble remembering or even distinguishing between the men who constituted the bankers and fund managers. They seemed to me to resemble harbor seals: sleek, smooth, and attired in shiny dark colored clothing. An occasional woman would turn up like a brightly feathered bird perched on a rock surrounded by seals.

American masters of the universe were somewhat more varied, even the bankers. There were always a number of young handsome men who traveled with us when we were on the road. They were mainly from Ivy League schools and dressed with expensive style. These were men who had made their way into the upper echelons of money and power well beyond the comprehension of most people. Their annual incomes were comfortably into the millions by the time they reached thirty. And that didn't count their bonuses. The real power rested with the senior staff, who never went on the road and more closely resembled the harbor seals in Canada, but would show up at celebratory dinners or at special lunches if they wanted to pitch some other kind of deal to their clients, people like us.

Once the investment bankers accepted us, they took us on the road to present our material to big portfolio management groups and pension funds. Investment bankers make their money when they sell your stock to such money managers. I suppose we told our story well, because the stock went up and we had no shortage of investment bankers wanting to do business with us.

I had thought that once we were making money we wouldn't have to go around talking to investors constantly. But I was wrong. We now had a lot of high-maintenance investors who expected to be updated in person at regular intervals by the management of their companies.

New York, Boston, and Chicago are the three cities where US money is concentrated, so this is where we spent a majority of our time when we were on the road. It was where our strongest shareholder base was. We also had to travel to Los Angeles, San Francisco, and Denver, where several big funds were located.

I had not appreciated how much time I would have to spend on investor relations, even after we became profitable. Between a quarter and a third of my time was now taken up traveling and talking to investors. I didn't mind the New York aspect of these duties, because that meant I could always take time to see Jennifer and Paul, to go to interesting restaurants they had discovered, and maybe get to the theater. Sometimes we could work our visits over a weekend, so Ed could arrange business travel so that we could meet and spend a couple of days together. These snatched moments were precious. Even when we did take holidays, we were always in contact with the company by phone or e-mail.

I remember one such occasion in the fall of 2000, when Elayne Wandler and I had started our journey with a visit to shareholders in Denver, where a large money management organization was a big shareholder. One of the men in the meeting room looked terribly ill and was coughing. I wondered to myself irritably why he hadn't stayed home. The next day Elayne and I flew to Boston for more meetings. We were to meet Ken in New York for our meetings there. During the flight my throat got scratchy and I coughed a bit. Then Elayne started coughing too. It was a very old plane we were flying on and we joked a bit about the possibility of toxic fumes from the engine.

But it was no joke the following morning when I woke up with all the symptoms of flu. I was running a fever, my joints ached, and I had a wracking cough. I met Elayne for breakfast and I saw that she too was ill. We had to get through the day somehow and then get to New York for a late afternoon meeting. I told Elayne I was going to get as strong medication as I could with no prescription so I could get my fever down and keep going. If I could get something to break the fever I'd be able to manage. Elayne told me she'd done a urine test that morning and found out she was pregnant, so she was afraid to take any medications.

It was a rough day. It is a blur of wretchedness in my mind. I remember we flew to New York for the final meeting of the day. The analyst we met with was a woman I had met before and liked. She was a scientist by training and had a dry British wit. She saw how unwell Elayne and I were. I said I was afraid we'd both got the flu while we were in Denver. She said, "You should get a prescription for Tamiflu. It's a drug from Gilead that's just been approved and it's supposed to work." I had read a bit about Tamiflu, a new approach to treating viral infections. The flu virus gets into susceptible cells in the body by attaching in a specific way to a receptor molecule on the surface of cells in the mucous membranes of the throat and lungs. The drug blocks the receptor so the virus can't get in. The drug works best if it is taken during the first twenty-four hours after symptoms start – the sooner the better, before the virus has a chance to spread. I said we didn't have access to a doctor to get a prescription and the woman very kindly offered to call her doctor and get a prescription sent to our hotel. Elayne, of course, said she couldn't take it.

But I had no compunctions. I don't think I have ever felt so ill as I did by the time we reached our hotel. The Tamiflu had arrived and I started taking it immediately. I was shivering and running a very high fever, as was Elayne. Ed and Anthony, Elayne's husband, arrived that evening to find their wives in a state of collapse. This was a Thursday evening. Fortunately Ken arrived and was able to deal with our Friday meetings in New York. Ed and Anthony acted as go-betweens for Elayne and me. I don't remember much about that Friday except that I was very ill. I can only imagine how ill Elayne was since she was taking no medication. I think I was even a bit delirious that day. I know I kept taking the recommended doses of Tamiflu and by evening my fever broke. By Saturday morning, my fever had gone down and I actually got up and went to see Elayne. She was really dreadfully ill, coughing horribly and still running a high fever and unable to eat anything. Anthony was so worried he considered taking her to the hospital.

By Sunday, my third day of Tamiflu, my fever had gone. I was able to get up and attend a theater matinee starring Judi Dench. I still had a wracking cough that persisted for several weeks. Elayne remained very ill and wasn't able to return to Vancouver until the end of the week. And after that she was still ill. The virus we contracted must have been a very potent strain. But I became a believer in Tamiflu and continue to advocate it. Elayne and my $n = 2$ experiment convinced me.

I have seen negative reports about Tamiflu but it seems from reading those reports that the drug is frequently misused. It is useless once the virus has spread throughout the lungs and is already in the susceptible cells. The critical aspect of the drug is that it has to be used as soon after the first symptoms of flu appear as possible. If treatment is not initiated during the first twenty-four hours of onset it won't prevent the viral progress. If I had started using it several hours before I did, my symptoms would probably have been less pronounced than they were. Jennifer and Paul arrived for Christmas a couple of years after this flu incident, both running fevers and coughing. They had become symptomatic soon after they boarded their flights. Tamiflu limited their disease in that instance too.

On another occasion, Elayne and I had a draconian schedule in New York and surrounding areas. We took the Cathay Pacific flight that goes from Hong Kong to New York via a stopover in Vancouver. It arrived in New York around 9:00 p.m. We had checked our luggage, because we were going to be on the road for several days and were carrying more than we normally did. We waited at the carousel until the last overstuffed wicker baskets and bundles were collected by our fellow travelers. My luggage did not arrive. Elayne's did. I had believed my luggage was safe because Cathay Pacific was a direct flight. It turned out that, for some reason, my luggage went to Hong Kong.

It was late and we had to be up at 5:00 a.m. to take a limo to Providence, a two-hour drive. We took a cab to our hotel. Elayne had booked us in at the "W," a trendy new hotel that was opposite the iconic Chelsea. I was wearing jeans. Elayne had an extra pair of dress pants I could borrow, but no extra jacket. At the hotel, we asked the concierge if there were any stores that would be open at this time of night where we could buy clothes. They couldn't help us but recommended a kind of drug store that carried some clothing. The concierge tried to be helpful and brought out an extra jacket that staff wore as a uniform. We set off and were able to find underwear and cosmetics. The hotel was indeed trendy. It was so dimly lit that you couldn't see to put on your makeup, and the halls were so dark it looked like there had been a power failure. But the jacket worked for me. The next day, when the limo brought us back to New York, I saw we had about half an hour before we were to appear on a TV interview with Bloomberg. I told the driver to stop outside Macy's on sixth and rushed in, purchased an outfit, and rushed out. My luggage turned up two weeks later in Vancouver.

Ciba was doing a good job both in North America and in Europe, where Visudyne had been launched. We had approval for Visudyne in Canada and were discovering the province-by-province struggle we had in getting healthcare reimbursement for the drug. This is the problem of the federal government delegating healthcare responsibility to the provinces. Retinal specialists tend to have group practices operating out of clinics and hospitals, so the healthcare system will pay for the drug if it is administered in a hospital. If it is administered in a doctor's office, the patient is on the hook for it unless they have private insurance. In the United States, since the vast majority of AMD patients are over the age of sixty-five, the national Medicare system pays for the drug and reimbursement only has to be negotiated once.

QLT had a very good year in 2000. Our quarterly sales continued to exceed expectations. Towards the end of the year we started getting letters of gratitude from patients who had received Visudyne treatment. They spoke about how our therapy had improved their lives. I was very moved by their messages and read some of them out at our Christmas party that year. I could tell that these testaments to what we had done moved everyone in the large ballroom. It was at moments like this that I truly appreciated the positive effect our research had had and felt so grateful to have had the opportunity to play a part in this journey.

At the time of peak sales of Visudyne we were treating about half a million patients a year. Wall Street might like those numbers because they translated into revenues. I liked them because they translated into the number of people we had helped keep from losing their vision.

CHAPTER TWENTY-FIVE

Growing the Company

The process of in-licensing or merger/acquisition in the biotech world of 2001 was precarious, difficult, expensive, and enormously time-consuming. We discovered this as we felt greater and greater urgency to grow QLT beyond Visudyne.

I realized that I was no longer the right person to be leading QLT through its next phase of maturation. My strengths lay in translating scientific discoveries into treatments and being able to shepherd them through development. I loved that aspect of the business – seeing bright ideas for treatments become a reality. I was far less keen on making deals for late-stage products and commercialization. And although I was beguiled by the way Wall Street had made us welcome, I was well aware that this enchanted situation could blow up just as fast as it had arisen. Also, the utter focus on money distressed me.

Our new friends on Wall Street would have no patience if we faltered. These new investors in QLT saw the potential of a product with annual sales of over half a billion dollars. If that prediction were not fulfilled, our stock would drop like a stone.

I was delighted by what we had accomplished but felt ready to hand the leadership over to someone with more business acumen. Ken Galbraith, our CFO, was a talented communicator, and he was very intelligent. He had already built a terrific rapport with the New York and Boston investment communities. However, he was not in the least interested in the day-to-day management of the company. Nevertheless, I felt that Ken along with the right kind of chief operating officer would make a superb management team that could take the company to the next level.

However, when I broached the subject with Ken, I was shocked to hear that he was planning to leave the company later in the year. We had talked about our futures and when we'd leave the company on several of our many road trips together, and he had said on one occasion he was considering leaving QLT after we got Visudyne approved; but I never really believed him, thinking he was too involved and too young to be able to walk away. Now I knew I had to believe him. He would be a wealthy man because he had many QLT stock options. He had a young family and he wanted more leisure. I was very saddened by his decision and realized it was now up to me to find a suitable successor as well as a new chief financial officer. Ken's unique skill sets would be very hard to replace and we would have a gaping hole in our management team. It was during my conversations with Ken that I had realized that a time would come when I would no longer be the right person to be CEO of QLT, once we were a profitable business. And I had mistakenly thought that he would be the person to step into the position after me.

Our next-generation photosensitizer, Lemuteporfin, was ready for clinical testing. It had a very similar clean product profile as Visudyne and was activated at the same wavelength of light. But we were not certain about where to use it. We looked at its potential in cancer treatments and did a thorough analysis of possible indications. We did not consider using it in the same way that Photofrin was used, as palliation for late-stage cancer patients.

Lemuteporfin's clean product profile would make it an ideal drug for treating early-stage cancers, where it would be curative. But market research showed us that such a treatment would never be accepted over surgery in cancer cases where surgery was considered potentially curative. With surgery, the tumor could be removed and the tissue examined microscopically to determine whether any residual tumor tissue had escaped the surgical margins. If it had, there could be further surgical intervention. With PDT, the tumor was ablated, leaving no residual material to examine. That kind of uncertainty would not be acceptable to patients or treating physicians. So we would be left with those rare cases of early cancers occurring in patients who, for health reasons, could not undergo surgery.

Then we hit upon the idea that we should look at the possibility of treating lesions that were considered pre-cancerous, similar to the use of Photofrin for Barrett's esophagus. We came up with benign prostate hyperplasia (BPH) as a possibility. This is a very common condition in older men, and although it is not necessarily a harbinger of cancer, enlargement of the organ itself may represent a small increased risk because of the larger number of prostate cells.

The prostate is a doughnut-shaped organ that surrounds the top of the male urethra, just before it enters the bladder. As men age, the prostate often enlarges, decreasing the width of the opening to the bladder. This can cause unpleasant symptoms like increased frequency of need to urinate and incomplete emptying of the bladder. Chronic pain and bladder infections may also be corollary complications.

Pharmacological treatments are many and not very effective. They function by relaxing the smooth muscle tissue of the bladder or blocking enzymes that may contribute to prostate enlargement. Treatment for more severe cases of BPH involves some type of surgical intervention to eliminate prostate tissue. There are three main procedures. TUMT, or transurethral microwave thermotherapy, involves the transurethral application of microwaves to produce sufficient heat to cause death of some of the prostate cells. Similarly, TUNA, or transurethral needle ablation, delivers heat to the prostate to cause tissue destruction. In most severe cases a surgical procedure is used called TURP – transurethral resection of the prostate – in which sections of the prostate are destroyed physically and removed via the urethra. These procedures are crude and are not always successful. In TURP, a device like a miniature bladed eggbeater is inserted into the urethra and used to mince adjoining prostate tissue, which is then suctioned out. Understandably, men do not relish the thought of this procedure.

We thought that patients who were in need of TUMT, TUNA, or TURP could well benefit from a more effective and possibly less unpleasant intervention using PDT. The market was sizeable, and we felt this indication presented an exciting possibility. But it would be challenging. We realized we had a large amount of preclinical research to do before we could possibly get into clinical trials for BPH. There was much we didn't know. How would we get the drug to the prostate? If we injected it directly into the prostate, how would we do this? If we could access the prostate, how would we tell how fast the drug disseminated in the tissue? It was a complex and intriguing possibility.

The interventional options open to BPH sufferers (TUMT, TUNA, and TURP) involved ablation or damage of the urethral tissue adjacent to the prostate, causing post-procedural discomfort, bleeding, and possible infection. If we could minimize these side effects, we could have a superior treatment. That meant we'd have to devise a way to get our drug into the prostate while avoiding its accumulation in the urethra.

QLT's device group, who were either engineers or physicists, took up the challenge and came up with an ingenious device for injecting the active drug directly into the prostate. It consisted of a balloon catheter into which was inserted a syringe with two retractable needles. When the syringe was positioned adjacent to the prostate in the urethra, the needles were deployed through the catheter and the urethral wall into the prostate. The drug could be injected, following which the needles would be retracted and the syringe removed. After an appropriate time during which the drug could diffuse throughout the prostate, a fiber optic could be inserted into the balloon catheter and red laser light shone through the surrounding tissue to activate the drug.

This ingenious design was evaluated by a focus group of urologists, who were uniformly supportive and indeed keen on the idea. We tested our approach in dogs and showed we could safely inject Lemuteporfin into prostates via the urethra with no ill effects. We were also able to treat the prostate with light and saw effective tissue ablation as a result.

We were fortunate to be able to conduct one more safety study that gave us full confidence that our approach was feasible. A group of surgeons affiliated with the prostate center at Vancouver Hospital were active in research and became interested in our study, and together we put together a protocol in which we recruited patients attending the prostate center for prostate cancer and who were scheduled for radical prostatectomy (complete surgical removal of the prostate). We asked them if they would allow us to inject our drug into their prostates prior to their surgery, using our device. Following surgery, the surgeons removed the cancerous area from the prostates and gave us the remainder of the tissue so that we could track the migration of the drug through the prostate. These men undergoing the surgery did not stand to benefit from what we were doing but gave permission of their own free will in the interests of medical research. I am always so impressed and humbled by people like these men who were facing a ghastly procedure and were willing to offer themselves up for the purposes of research.

It took months to carry out all this work. We knew we wouldn't be ready to treat BPH for more than a year when we started on this project, but the promise that our treatment might offer was considered worth the effort.

The BPH project really got my attention. Solving the complex problems around getting the drug to the targeted tissue and avoiding discomfort to the patient was challenging. I enjoyed discussing ideas with the scientific staff, raising challenges and trying to come up with clever solutions. I realized how much I missed my association with scientific thinkers and how much more I liked being with them than with bankers. I wondered if, when I found a successor, there still might be a place for me at QLT dealing with scientific issues.

That year, 2001, I started looking hard for a successor. It became common knowledge around QLT that we had engaged an executive search group to find candidates to replace me. I was touched by how many people tried to talk me out of this step and urged me, if I insisted on leaving my position, to find someone who wouldn't change the corporate culture. I was gratified to hear this, realizing I had played a part in creating a workplace where people were happy. But when I thought more about this, I realized that people who liked the kind of corporate culture we had would naturally want to work here by a process of natural selection. Those who wanted something else would leave.

But I took the request seriously and discussed it at length with Linda Lupini, our vice president of human resources. She suggested that I undergo a personality analysis so we could look for someone with similar characteristics. She brought in management consultants who administered the tests, which consisted of hours of questions, multiple choice in the main, that seemed incredibly repetitive but had subtly nuanced differences between similar questions. The reams of answers were analyzed and in the end the findings were put into charts that were supposed to profile my basic personality and management style. I've always been fascinated by these kinds of tests, even though

I have doubts about their ability to really define a person's character. When the analyses were complete one of the senior people from the consulting group came to go over the findings with me.

We sat together and went through the charts. For the most part there were no surprises, until we came to one chart. It looked rather like an airplane propeller. It was supposed to yield something resembling a circle with some skewing depending on individual personalities and penchants. The horizontal and vertical axes started at 0 per cent at the center where they intersected, and went to 100 per cent at the extremities. Personality traits were listed around the circumference of the circle at the 100 per cent limits. Traits like visioning, problem solving, a leaning to leadership, or a leaning to following were the main ones at the extremities of the vertical and horizontal axes. The tips of my "propeller" that indicated close to 100 per cent designated visioning and problem solving; the nadirs that were close to 0 were inclination to leadership or followership. The consultant looked at me.

"You're never going to find someone with your personality who wants to be a CEO," she said. She went further to say I was introverted, that I lacked self-confidence, and that I wasn't concerned about my lack of confidence. I had a profile characteristic of an academic scientist. Nothing she said surprised me. I know myself quite well and have some affection for my inner demons. They keep me grounded. Being the CEO of a profitable company had never been part of my career plan and I knew, although I wouldn't admit it to myself, that I really didn't like being the CEO of what QLT had become. The consultant went on to say that no one with this kind of profile should ever be considered for a CEO position. She suggested we just "bury" this result.

Regardless of the problem of finding a CEO who had no desire to be a boss, we went ahead with our search. Over the next few months I learned a great deal about people who wanted to be CEOs, and more often than not was singularly unimpressed.

That summer we identified a possible in-licensing opportunity. It was a chemical that was being developed as a possible adjunctive treatment for cancer patients. The product was owned by Xenova, a British biotech company that was having financial difficulties and had a couple of products in phase II clinical trials. They didn't have sufficient funds to continue to finance the development costs of all their potential products. The product they wanted to out-license was called Tariquidar, and it was a P-glycoprotein inhibitor. P-glycoprotein (P-gp) is a protein molecule occurring on the surface of many cells in the body. P-gp functions as a pump that increases outflow from cells of waste or possible toxic substances. P-gp is present in high concentrations of cells in the body that may be in contact with toxic materials, like the gut, kidneys, liver, lung, and blood-brain barrier. I think of P-gp as a kind of toilet or drain that flushes away unwanted materials in cells.

Some cancer cells can mutate so that they produce increased levels of P-gp on their surfaces, and become multi-drug resistant because they flush chemotherapeutic drugs out as fast as they enter. These mutated cancer cells can very efficiently eliminate cytotoxic drugs that are used in treatment. This can occur in breast, lung, and kidney cancers. Tariquidar is a molecule that plugs up the flushing ability of cell-surface P-gp and thus renders cancer cells more susceptible to chemotherapeutic agents.

Ed and I met with one of Xenova's senior business development people when we were at BIO, a huge international biotech conference that is set up to enable companies to network and present information. That first meeting went very well and we thought the Xenova people would be reasonable to deal with. They had some early phase II data that convinced us that the drug had the activity of a P-gp inhibitor.

We were able to come to a mutually satisfying arrangement with Xenova that summer, our first successful major in-licensing agreement. I remember the final negotiations being rather tense. It was July, and Ed and I had managed to clear a few days to get up to our island retreat on Sonora. One of Ed's senior staff and one of our legal staff were handling the negotiations in the United Kingdom. We have had telephone woes at Sonora since we first went there. Even now, in 2017, we still don't have any reliable phones on the island, although we have had reliable satellite high-speed internet since 1999. We got an urgent e-mail message from Celia Courchene and Modestus Obochi, our negotiators in London, saying that Ed had to contact them to get to solutions on the final agreement. There were a couple of tough questions they were grappling with. It was about three in the afternoon when the message came through and that meant late evening in London, so we had to get back to them as soon as possible.

Owen Bay, where our property is, is surrounded by tidal rapids. On one side is a tidal flow called the Hole in the Wall, where Sonora Island and Maurelle Island almost touch. This narrow passage accommodates an enormous tidal flow through it every six hours and creates overfalls and boiling whirlpools. Boats usually wait and go through at slack tide. On the other side of Owen Bay are the Okisollo Rapids, not quite as vicious as the Hole in the Wall but bad enough to deter most boaters except at dead slack tide. These natural filters were in part what attracted us to the property.

We knew we had good phone reception out in the rapids, so we took our boat, a sturdy welded aluminum craft especially designed for coastal waters, and went out to the rapids to make the call to our colleagues in Britain. It was a beautiful day and the rapids weren't running too hard. Ed was the head of business development, so the call was to him rather than me. I had the responsibility of keeping our boat safe while he was on the phone with the negotiators. We killed the motor so Ed could hear better and I just sat at the wheel and watched to see when we drifted too close to a whirlpool or were in danger of getting close to the whitewater. When that happened I had to start the engine and get us to a calmer location. At one point we got caught in the outer edge of a massive whirlpool and started drifting around in circles. This was a fair incentive to reach a conclusion.

The conversation seemed to have done the trick, because we ended up with an agreement on Tariquidar that summer of 2001. We agreed to pay ten million dollars up front and agreed to fund the phase III clinical trial for Tariquidar. QLT would retain marketing rights for the United States, Canada, and Mexico, and Xenova would market in Europe. The product itself excited me. If it were successful in clinical trials it would be a very important addition to cancer therapy. We had a good clinical team. Mohammad, our head of clinical, had built a strong department of competent people and he was an established oncologist.

Now we just had to deliver good results, in trials we had to design.

CHAPTER TWENTY-SIX

Time to Reflect

In 2001, the week after 9/11, while the world was still in shock from Al-Qaeda's attack on the World Trade Center, we had planned to fly to Switzerland for our quarterly meetings with Ciba Vision. But schedules were completely disrupted for days following the attacks, and airlines were struggling to accommodate travelers and reopen airports with new security measures and a new batch of security people in place.

It wasn't clear that we'd be able to get a flight. We could have held the meeting by videoconferencing but there were issues over marketing plans that Lee Anne and Bob Butchovsky, our marketing director in charge of Visudyne, were keen on discussing face to face. Ed and I had planned a brief holiday in Tuscany following the meeting and decided that if we could get there, we would.

Air travel out of Canadian and American airports was still chaotic but we were able to obtain tickets for Zurich flying out of Toronto. The journey was arduous and long. The Toronto airport was packed with irritable people. Lineups at security had wait times of three hours or longer as airports rushed to try to get new security measures operative. Everybody was on edge. Newly hired security people were untutored and apt to confiscate anything that looked like a possible weapon. I saw the woman ahead of me in line have her eyelash curler confiscated. She was speechless when the security guard inspected it and threw it in a bulging bag at his feet containing nail files, nail scissors, and other confiscated items.

The Zurich airport was an oasis of calm in comparison to Toronto. Nothing seemed different from our previous arrivals. If anything, the airport was quieter than usual. The commuter train to Basel was on time. But here also, people seemed subdued.

I think people everywhere were shocked and silenced by the enormity of what a handful of terrorists had succeeded in doing. Nobody had ever considered using a plane full of people as a weapon of mass destruction before. Now, no one would ever forget.

During our meetings we heard, for the first time since launching Visudyne over a year before, that the sales people at Ciba had started to notice a slight slowing of sales. That wasn't surprising. It seemed logical that at some point sales had to top out, unless the market size was increased by approvals in more jurisdictions. There had been a backlog of people needing treatment when we launched. Now, those patients who had been waiting were in treatment. August was always slow in Europe. But there was clearly concern within the sales force. I had to prepare to tell the investment community that projections were likely to change.

Ed and I spent ten days in Tuscany after the meeting. Cities like Florence felt empty now that they were devoid of tourists. And for a few days I was able to ignore the concerns raised at the Ciba meetings. But I had anxious phone calls from Bob, who was worried about sales projections and was keeping daily checks on how Ciba was doing. I returned to QLT with a feeling of unease. September 30 was the end of our third quarter, and if sales didn't pick up we would miss our EPS forecast. Surely people would understand, I argued to myself. 9/11 had thrown a long shadow over everyone. Even people who might need a treatment could have many reasons for canceling. As it turned out, we just squeaked by with our sales forecast and our EPS satisfied the market. But I wondered what the next quarter would bring. Again I realized how much I did not like being CEO. I challenged myself. Did I want to leave my position because things were getting tough? My expertise was research and development. I quickly answered myself – no, I wasn't a quitter. I had spent years doing basic research where failure of experiments is the norm. I'd never wanted to quit then. I knew I loved the company I'd played a part in creating and the people in it. I had helped it become a very successful Canadian biotech company with a market capitalization of several billion dollars. I was proud of that and wanted us to continue to do well. I could see we were entering a new period where the knowledge and experience needed were not skills I possessed. Marketing and corporate development were areas in which I was very uncomfortable. It was definitely time for me to find a successor.

I pressed our headhunters to find appropriate candidates for my job. I had already been through a number of interviews with possible candidates who had looked fine on paper but had turned out to be very disappointing.

Over the next five months I saw the resumes of over forty candidates. I found myself continually surprised at the kind of people who applied. I had phone conversations with possible candidates whose resumes looked appropriate. If the conversations seemed promising, I would make arrangements to meet by bringing them to QLT, or meet them at the various investor conferences that were held all fall in New York and San Francisco.

I met some enormous egos and people who had the confidence to think they could walk into a very successful company with a seasoned and capable management and just take over, without wondering for a moment if they would be a good fit or demonstrating that they understood the challenges that lay ahead. QLT was a star in the biotech firmament and I felt that a respectful recognition of that fact was appropriate. I also expected candidates to be informed enough to ask probing questions about the company, about the challenges we saw, and what I thought were the most pressing needs of the organization. I'm still amazed that so many people have such confidence in themselves that they don't feel it necessary to understand the kind of environment they're trying to become a part of. I was looking for someone thoughtful and questioning about whether they'd be a good fit, looking to find something out about the company they would be leading

I asked myself if I was being too critical, and whether in reality I didn't want to give up my leadership. But that kind of soul-searching only led me to a fervent desire to find the right person. I definitely wanted to hand over leadership. I knew it was to be expected that any change in top management would cause uncertainty and unhappiness on the part of some of the employees, and that troubled me. I felt such enormous loyalty to the people I had worked with for the past years that I was fearful of making a decision that would result in unhappiness and dysfunction at QLT.

I think that part of my fearfulness about hiring the wrong person was due to my experiences during the years that Jim Miller was our CEO. His habit of hiring people he'd met on a plane, or in some other unlikely way, into senior positions taught me how much grief could result from hiring the wrong person for a position of authority. I also knew how difficult it was to get rid of someone once they were hired into management.

From the interviews I did at this time, I saw how people betray themselves in small ways: the way they refer to their own staff, amusing anecdotes about themselves that reveal a character slightly less than honest, or their exclusive use of the first person when referring to accomplishments that must have involved the cooperation of others.

I think the most disappointing interview I had was with a woman. Out of the forty plus candidates I'd looked at, she was the only female who applied. And she was well qualified. I had been delighted to see her resume come across my desk. She was an MD-PhD and was in charge of a major business unit of a biopharmaceutical public company in the US. Her credentials looked first class. We brought her to QLT to interview. The candidate was a small, bird-like person with a tense face. I stood at my office door to welcome her as she strode purposefully down the hall in front of Susan, my executive assistant, who was showing her the way to my office. The candidate and I shook hands, after which she took of her coat and more or less threw it at Susan without giving her a glance. As we went into my office and shut the door I saw Susan's face. Her eyes were saucers and she was flushed. Susan is an incredibly professional person and I don't think I have ever seen such a look of discomposure on her face before or since.

It wasn't really an interview. She and I spent an hour with me listening to a monologue of this woman's accomplishments. She seemed completely incurious about the sort of company we were. I tried to lead her into conversations about QLT, to probe how she would approach the possibility of becoming the CEO of an existing profitable enterprise, but she seemed disinclined to talk about anything other than herself. I asked her how she felt about the possibility of moving to Canada. She looked surprised for a moment and said the thought hadn't crossed her mind. She'd assumed there would be no difference.

When we came out of my office, Susan still had that look of horror on her face. After I had guided the candidate to the washroom I looked at Susan and wrinkled my nose and shook my head. Susan's face broke into a smile. "That's all I wanted to know," she said. "I've been sitting out here for the last hour almost in tears."

I had to go to New York in November to attend one of the big institutional meetings. I would have the opportunity there to interview a number of possible CEOs. The meeting was held at the Plaza, an iconic New York hotel on Fifty-Seventh and Fifth Avenue. I hadn't been in New York since 9/11, and Elayne and I felt nervous about going. Jennifer and Paul had told Ed and me how different the city was since the attacks. So many of the fund managers and bankers we would be meeting with were headquartered either in the World Trade Center or in its immediate surroundings. I wondered how they had fared.

This would be Elayne's last investor relations activity before she went off on maternity leave. Ken, our chief financial officer, had left the company in late summer, and we both missed him as we traveled to New York. The three of us had spent a lot of time together on the road and had got along well. It seemed like everything was in a state of flux.

New York was not itself; it seemed somber and slower paced. Of all the cities I have visited, I have always felt that New York has its own unique energy and vigor. You feel it as soon as you get off a plane. That feeling of high energy had been dampened. The city was empty.

We met with many analysts and bankers at the conference. 9/11 was still on everybody's mind and the subject surfaced in most conversations. Understandably, people seemed keen to tell their personal experiences. I talked to one of our analysts whose business was located about a block from the twin towers. Their building had been damaged and had been vacated for ongoing repairs. He told me about how he felt that day, the smoke and the fear that was so pervasive and remained so for days. He said he was okay, but I could see that he wasn't. In retrospect I think many of those people who were carrying on with their lives as normal after this traumatic event were suffering from PTSD. There was one fund manager, a young woman I had always liked for her low-key questions that held great depth, who told me of a conversation she had with an aunt who was trapped in her office in one of the towers. The aunt knew she was going to die. The fund manager said she talked to her aunt until the phone went dead. The niece's offices were uptown, so she had not been at risk. She was suffering from survivor's guilt. Her face looked tortured.

Usually, these conferences organize evening events in the form of theater tickets, dinners, or concerts. More often than not, in the past, there were fund managers and bankers ready with invitations to expensive restaurants. This time, those invitations were not forthcoming and we found time on our hands in the evenings.

One evening, Elayne and I took a cab down to ground zero. It was a bit frosty that night and clear. The smell from the devastation was still all over New York at that time. You became aware of its smoky oily presence as soon as you landed at La Guardia. As we approached ground zero, the smell became toxic. I thought about all those people working there day and night in that atmosphere and hoped they were protecting themselves.

We followed a number of other figures moving closer to what we guessed was ground zero. Ground zero was then and continues to be an eerie place for me. At night it created the impression of being some kind of inferno, with lights hooked up around where people were working around the clock clearing the rubble, and the endless release of dust and debris into the air making it hardly breathable. The sounds of heavy machinery creaking and groaning added to the hellish effects. We stood on the catwalk that had been erected for observers and said nothing. There was nothing to say.

I carried out more frustrating interviews while I was in New York, but came back empty-handed. Having lost Ken and now with Elayne going off on maternity leave, I wasn't feeling very confident about the future.

Visudyne sales were disappointing for October and I was dreading the year-end. No one was willing to suggest that 9/11 was affecting, directly or indirectly, the number of people seeking ocular examination for vision problems. And whether it was or not, Wall Street was only interested in the bottom line.

CHAPTER TWENTY-SEVEN

An Enormous Rush of Relief

Early in January in 2002, Bob Butchovsky came to me with the news that Luzi von Bidder, the CEO of Ciba Vision, had announced at a meeting in Europe that Ciba Vision had not reached its forecast for sales of Visudyne in the fourth quarter. They had missed their target by about four million dollars, not a large number when sales were expected to be in the range of fifty to sixty million.

I knew that sales had been slow during the quarter, particularly in the United States, but I had hoped the expected number would be achieved with sales in Europe. Now I was in a quandary. Did I need to issue a press release announcing that we had missed our numbers, or could I wait and provide the information in our quarterly report?

Luzi had revealed his estimates of the quarterly sales at a meeting that, although not public, had been attended by a large number of people at Novartis. The people at the meeting had not been told that the information was confidential. It was possible that some analysts might get hold of the information.

I had a legal obligation to make public any material event regarding QLT. A material event is, broadly speaking, any event that might affect our stock price, either adversely or otherwise. And this information no doubt was going to have an impact on our stock price.

My question was, should or could I hold off on reporting Luzi's numbers until we reported our quarterly results about ten days from then, after the numbers had been fully integrated into our financial report? Making a bald announcement about falling short of our projections would likely affect our stock price more adversely than a carefully nuanced announcement that presented this information along with other significant events for the company.

But to wait would risk someone finding out and spreading the information. I wished Ken were still with the company to provide advice. But the only person we had in finance was our controller, a competent woman who had come from a private company and had no experience with public companies. I was still looking for a CFO. And I wished Elayne were still with us. She had left the previous month on maternity leave and had hired a young man to replace her, who had come from the sales force of Amgen Canada. Ian Harper had some investor relations experience, but he was green.

I made the difficult decision to announce the fact that we knew that we hadn't reached our predicted sales numbers. I couldn't countenance the thought that we would go on trading our stock with this knowledge possibly in the hands of some fund managers, and not report it.

We put together a press release providing the information and sent it out after the close of the market. The stock market in North America is synchronized with the New York Stock Exchange, so market close is at 4:00 p.m. New York time, 1:00 p.m. on the west coast. Similarly, the market opens at 9:30 a.m. New York time. I was up after a bad night and at QLT at 5:00 a.m. with Ian the following morning, getting ready for the call that would take place at 6:00.

I made a brief statement and took questions for a short while. There were a lot of people on the call but there weren't a lot of questions. We had disappointed Wall Street. Then we sat and watched the price of our stock take a nosedive. Our stock at the time was trading around sixty-five dollars Canadian. That day we lost 30 per cent of our value and ended up in the mid-forties. A huge number of shares changed hands that day, over ten million. Our staff in investor relations had a rough day handling calls from irate shareholders. On days like that you can only cheer yourself with the knowledge that people are buying stock at these reduced prices, because you can't sell a share without having a buyer.

It was a pretty glum day around QLT. Wall Street had spoken. We hadn't measured up. I tried to make sense of this massive reaction to the missed numbers, but I couldn't. The previous quarter had been a tough one, with 9/11 throwing the world into turmoil. The only way I could make sense of it was that the stock market had reacted this way to the missed numbers because they were an indication that Visudyne might not be as big a product as we and our investors had hoped.

We had to carry on, do our best to build the company, and put this awful day behind us. The following week, one of our legal staff came to me with the news that a class action suit had been filed against the company by a US law firm known for class action lawsuits. I had heard of these kind of lawsuits and had always thought that they were filed when management was caught defrauding the public, like lying about gold findings or sales numbers. I was stunned. We hadn't done anything wrong. We hadn't lied. And no one had traded shares since we heard the sales numbers. Insiders were barred from trading in the weeks leading up to reporting quarterly results.

The basis for the lawsuit, it turned out, was that when Ken had left the company the previous fall, he had exercised all his stock options and sold the stock. At about the same time, I and a few other people at QLT realized we had a sizeable number of stock options that were going to expire by year end, and so decided to exercise at that time. Up to this point, since we formed the company, I had never sold any shares or exercised any options, though Ed and I had given some shares to our children. I was sensitive to the fact that as CEO, my selling shares would not be taken lightly by shareholders. Sales of shares by executives in publicly traded companies are carefully scrutinized by the investment community. I had exercised these options because they would expire and be valueless in a couple of months.

By far the best time for executives of a public company to sell shares is almost immediately after they report their quarterly results. At that time the management is duty-bound to report all material events in the company, so immediately after that reporting is done, it is assumed that there are no unreported material events that insiders know about. These were the only shares that I sold while I was CEO, and although this sale made our family secure, the options I exercised at that time represented a very small percent of the shares I held. Now we were being sued because the claimants asserted that back in September we knew the sales projections for Visudyne were too high.

I came to understand another, seamy side of money markets and the kind of people who take morally questionable opportunities. The New York–based law firm that was coming after us existed for the purpose of mounting these kinds of class action suits. They followed the market and watched for sudden drops in the share price of any company trading on the exchange. They research the reason for the drop to determine if any people from management had been selling shares in the previous quarter, and if they have, these lawyers go after them on the assumption that the employees had prior knowledge of some coming adverse event and sold shares prior to its announcement. Ken and I and several other people in management were named. I was horrified and indignant. Our legal people were confident that the suit would be dismissed, because we had done nothing wrong.

In order to mount a successful class action, the lawyers had to sign up a number of angry shareholders who thought the company had deceived them. It turned out that these lawyers had problems with their suit against us because they weren't able to enlist any significant shareholders. Nevertheless, they went ahead with the suit, which we eventually won with prejudice to the lawyers bringing suit.

But I felt soiled by this experience and longed to be free of my position. The suit itself took several months and bringing it to conclusion was very costly to the company.

Around this time I also became very disheartened about finding my successor. I had moments when I wondered who would want to become the CEO of QLT? I no doubt had a warped view of our company at that moment, seeing it as a failure. For anyone looking at us from the outside, we were still a very successful company. The bruises we had received from Wall Street were not visible. We had an excellent product that was making us very profitable. Visudyne sales continued to be brisk, although they were lower than Wall Street had expected. On reflection, I realize that most analysts who followed our stock had predicted the Visudyne sales would earn in excess of half a billion dollars a year at peak times. Our stock price was based on that number. It turned out that in the following years Visudyne sales didn't quite reach that number, with sales earning between four hundred and four hundred and fifty million dollars a year.

It was in the spring of 2002 that a resume came across my desk of someone applying for the CEO position. There was something about the brief resume that got my attention. The applicant had a bachelor's degree in pharmaceutical sciences and had been in bio-pharmaceuticals for about twenty years. His job experiences were in roughly five-year spans in four different companies. What got my attention was that he left the companies he was employed by in order to take on positions with an increased level of responsibility and seniority. Now he was CEO of Celera, a San Francisco–based biotech company. I checked out the company and discovered that it was in the process of being sold. That explained his interest in the position at QLT. This career trajectory seemed to me so clear-cut and unequivocal that I assumed this person really knew where he was going and what he wanted. Not only that, he appeared to have suc-ceeded in achieving his goals at every new position he took.

I called Linda Lupini, our vice president of human resources, and asked her to come to my office so we could talk about the resume I had seen. She hadn't seen it, which was surprising since the head-hunters usually sent resumes to both of us. The candidate's name was Paul Hastings. Linda agreed that this resume was worth follow-ing up and said she'd arrange with the headhunters for an interview.

I learned later that I had received this resume in error, since the headhunters never intended to send it to me. When I asked why I was told that they had assumed that because our company had so many PhDs in it we wouldn't be interested in a candidate who didn't have that advanced qualification or an MD. But in fact, we wanted a CEO who could achieve the goals we set as a company. A PhD or another kind of post-secondary degree would not be a guarantee of that. I was curious about this applicant. We arranged to meet with Paul for an initial interview in San Francisco, where I had to be the following week. Linda would meet me there.

My first impression of Paul, when Linda and I were shown into his office, was that of a grounded person who didn't feel the need to impress. A clean-cut man in his forties, Paul exuded a quiet con-fidence that I liked. He had an open square face with a nice smile and close-cropped slightly receding blond hair. He was very casu-ally dressed in cotton pants, a T-shirt, and a V-necked sweater. He was clearly someone who wasn't out to impress us with his dress code. I assumed he was telling us who he was. He didn't waste time in small talk and proceeded to ask a series of probing and coherent questions about QLT. He asked what our vision for the company was. I was delighted. He was the first person we had interviewed who had done this. He was someone who really wanted to know what QLT was about.

He didn't hold back about himself either. Over the course of our conversation, Linda and I learned that he was gay, had a partner, and that he'd had Crohn's disease since he was a teenager and had undergone colostomy surgery. He was at this time putting the finishing touches on the sale of the company where he was currently CEO. He spoke succinctly and briefly about his prior experiences without any self-aggrandizement.

We agreed to bring him to Vancouver so he could meet our management and see the company. Linda and I both had the same impression of him. We both liked him and had found him open and easy to talk to, with an ego that was healthy but well under control. We checked his references and got glowing reports back from his previous bosses.

He came to Vancouver the following week, where Linda and I had dinner with him. In the casual setting of sharing a meal, I discovered that Paul had a good sense of humor, was actively involved in raising money for a camp for children with Crohn's disease, was very open, and was a Democrat. I also discovered that he had not been idle since we had seen him last, as he presented us with another long list of questions about the company. He was the first candidate I had met who I felt could take over QLT and bring it to another stage of development. Most of his experience had been with companies that had products on the market, and some of his activities had involved sales and marketing. Although the commercial aspects of running the company were familiar to him, he also exhibited a genuine interest in the science behind Visudyne and Tariquidar.

I woke up the next day feeling hopeful, and then had an awful thought – maybe he wasn't really interested. I called him at the hotel and was very direct with him. I asked him straight out if he was serious about coming to QLT. I didn't want to waste time if he wasn't serious. He assured me he was very serious and that he'd been worried that I wasn't. At that moment I felt an enormous rush of relief.

I made sure he met all our management team and was relieved that there was general approval on their part. The next task was getting approval from our board. Duff Scott, as chairman of our board, came to Vancouver and met Paul and was happy to recommend him to the rest of the board.

Paul became CEO in the spring of 2002. He had asked me if I wanted to stay on in some capacity when he took over. I admitted to him that I was interested; I would be happy to help him in the science end of the company's business, such as doing due diligence on possible in-licensing opportunities and participating in some way with the company's ongoing science projects. Paul was very happy with this possibility and suggested I take a position as executive chair of the company's scientific advisory board, so that on paper I wouldn't be reporting to him.

When Paul called Duff with his suggestion, Duff was very negative. He came from the old school that believed that ex-CEOs should not be involved with the company once a new CEO came in. I was incensed by this attitude, but when I thought about the people I had interviewed for the CEO position I realized that many people who aspire to this post would find it difficult to stand by while someone else picked up the reins. Not me – I was only too happy to hand the reins to someone I felt could do a better job in growing the company.

Paul went to bat for me over this issue and won the day. I moved out of the corner office happily and looked forward to helping Paul and the company in the best way I could.

CHAPTER TWENTY-EIGHT

A Biotech Company with a Lot of Money Is in Danger

I was back doing exactly what I wanted to do. I believed I could provide useful advice and analysis to the company and to Paul. We set up a group to select the next-generation photosensitizer to follow verteporfin. David Dolphin's team of chemists got busy and some of our scientists began evaluating the new molecules being generated in David's lab.

I became part of the team that was following the prostate hyperplasia studies. I also became involved in the due diligence process for evaluating possible in-licensing opportunities. I was available to travel to other companies whose products we were considering licensing.

Paul and I continued to get on very well. We liked each other. He was a forceful man. He readily extended his friendship to those around him and was the kind of person who faced the world openly, expecting the best from people he worked with in return. Also, he was very smart. The one person in senior management who felt completely at odds with him was Lee Anne. Perhaps it was the fact that Paul's experience in marketing rivaled Lee Anne's and she didn't like the competition. Whatever her reasons, she decided to leave the company. She had built a significant marketing team around her and had been a tough and demanding boss. I know some of them were relieved by her departure. But I missed her. She had provided me with solid and reliable advice in some pretty rough times.

We completed the preclinical work on benign prostate hyperplasia (BPH) and applied to Health Canada and the FDA for permission to go into the clinic with the procedure. I was excited. Our engineers and other scientists had put a lot of thought and expertise into developing the device to be used and doing the preclinical research. We had known the preclinical work would take a long time, but it had gone smoothly considering the complexity of the project. We decided to use a standard dose of the drug with escalating amounts of light treatment. The way studies like this are done is that the first three patients receive the lowest light dose. Then there is a wait period of a week or two until follow-up to ensure that no adverse events occur after the treatment. After this time period, the treating physician will escalate the light dose in the next cohort of patients in a stepwise manner. The first good news came back after the first cohort of patients was treated. They had tolerated the treatment very well. There were no reports of major discomfort either during or after treatment.

The follow-up after treatment involves the patient returning to the doctor's office over a three-month period to report on symptom improvement. BPH is a condition that physicians cannot monitor by examining the organ itself. The only assessment that can be quantified is the rate of urine flow, which can be measured with some accuracy. Prostate size can be determined through palpation, but that may have nothing to do with symptom improvement because symptoms are caused by constriction of urine flow from the bladder through the urethra, something that would not be revealed by the external diameter of the prostate. Therefore the patient is the only really valid evaluator of symptoms. The most troublesome symptom of BPH is a sensation of the need to urinate frequently, particularly during the night. So this, as well as other lesser symptoms, are evaluated in determining if a new treatment is having an effect. We were warned that in trials for this condition the placebo effect could be significant.

Early on in this phase I study, we were getting reports of symptom relief from both physicians and patients. And we were hearing that patients were not suffering any significant unpleasant side effects. Our hopes were high that we had a new therapy for this condition. The doctors loved it too, because it involved a procedure they could do in their offices and charge for, rather than prescribing a pill.

That summer, after a great deal of internal debate, we decided to push forward with a phase III trial for Tariquidar, our P-gp inhibitor. Xenova was happy with our decision. The decision on how to proceed with Tariquidar was reached after much argument. I attended most of these meetings, which were largely run by Mohammad, our chief medical officer. He was a trained oncologist and his forte when he was at Sanofi had been phase III trials. It was natural, then, that his voice was very persuasive. There had been a group of people, myself included, who had opted for caution. While Xenova had done some phase II work with the drug, there were holes in the data that left a number of uncertainties when putting together the protocol. We all knew that the drug was a very efficient inhibitor of the cell surface pump (P-gp). But just how much of the P-gp inhibitor to use in combination with some of the chemotherapy drugs commonly prescribed was not known, so there were safety concerns and risks. More cautious voices, including mine, suggested running a couple of small phase II studies to clarify these points. Regardless, the decision was made to go ahead with a very large phase III clinical trial for non-small-cell lung cancer in patients diagnosed with advanced disease. Xenova and Mohammad were both keen on taking this path. The first-line therapy for these patients was either a combination of Paclitaxel and Carboplatin or Vinorelbine on its own, depending on standards of care in various jurisdictions. In addition to the chemotherapeutic drugs, patients in the trial would receive either Tariquidar or a placebo. Overall survival would be the primary end-point, although other parameters like tumor shrinkage and toxicity would be assessed also.

The trials would be run at over twenty sites in North America and Europe. For the first time, we included sites set up in countries that had formed the former Soviet bloc. This was partly because lung cancer incidence was much higher in these countries than it was in Western Europe, and partly because there was less competition for patients to enroll in trials in those countries.

Paul, our new CEO, was experiencing some growing pains in dealing with our analysts and shareholders. He had come from being the CEO of the biotech company Celera, and he had overseen its sale to Applied Biosystems, a sister company owned by a third organization, Applera. He had only been CEO since the departure

of the company's previous CEO, Dr. J. Craig Venter, Celera's re-
nowned sequencer of the human genome. Venter had become
famous beyond the biotech sphere because his company, Celera,
had been first to sequence the complete human genome, ahead of
the multi-investigator, multi-sited effort funded by international
governments. Paul had not had to face analysts and investors,
because Venter would have handled all those kinds of activities as
the public face of Celera. But Paul learned fast after being savaged
a couple of times.

He realized that he needed someone other than a staff member
from investor relations accompanying him when he traveled to talk
to investors. Mohammad had recently earned his MBA from the
Ivey Business School in Ontario. He had discussed his plan with
me a couple of years earlier when he enrolled in the program. He
had aspirations beyond chief medical officer, and made it clear that
he wanted to become a CEO at some point. Paul knew this, and
suggested that Mohammad accompany him at investor meetings to
provide details and explanations of scientific and clinical activities.
Mohammad was keen to extend his experience in that way. While
Mohammad was not inclined towards basic science, he was a fast
learner and had a head for detail. And his knowledge of clinical
details was prodigious.

With Mohammad often engaged in corporate affairs, the company needed another experienced clinical person. Paul hired a man
named Mo Wolin, whom he had known from Chiron, the biotech
company where Paul had been a vice president before moving to
Celera. Mo was a trained oncologist but his real passion was for
molecular biology. I was pleased with this appointment because Mo
had a deep interest in science, which I felt was important in planning early-stage clinical trials. Mo was young, quiet-spoken, and
had a strongly academic approach to his work. I enjoyed talking to
him about his ideas. He had the air of an absent-minded professor,
which I found refreshing.

We were making a lot of money from Visudyne sales. A biotechnology company with a lot of cash is in danger of losing control to
activist shareholders, who can take over the board of the company,
use the money to pay dividends to shareholders, and basically strip
the company financially. We had noticed that blocks of our shares

were being sold to some questionable hedge funds out of New York and took this as a danger signal. Paul redoubled our efforts to find a suitable acquisition.

Even though Visudyne sales were continuing to do well, now for the first time we had potential competitors on the scene. One of these was a product called Macugen, and the company taking it forward was a Florida-based company called Eyetech. Macugen was an anti-VEGF product that was being injected intravitreally directly into the eyes of patients every six weeks. Eyetech was carrying out two pivotal studies on patients with wet AMD. End points were similar to what ours had been. In a similar time frame, Genentech announced it was also running a phase II study with their anti-VEGF monoclonal antibody, Lucentis. They also were injecting their drug directly into the eye. We couldn't believe that patients would opt for a therapy that necessitated intra-ocular injections at that kind of frequency. Still, we had to track this potential competition carefully.

We looked seriously into at least half a dozen possible acquisitions during 2002 and into 2003. This kind of activity takes an enormous effort on the part of the acquiring company. There is due diligence to be done, which involves legal, scientific, financial, and clinical expertise. Then there are the almost endless meetings and discussions with management from the potential acquisition. I was impressed that in the companies we looked at, there were so many people in senior management who cared about what a merger would look like for their employees. They wanted to know if employees in the acquired company would continue at their current level of seniority. Then there was the inevitable question as to the determination of what the acquisition target was worth. And what value would share options and shares in the acquired company be worth post-acquisition?

CEO egos were often the cause of deal breaking. Someone wrote once that a CEO would probably not be a fun date. Whoever wrote that was right, with very few exceptions.

I remained on the board of QLT after Paul's arrival, so I had the opportunity to assess their reaction to our new CEO. I got the feeling that many of the board members felt distanced from Paul and that they didn't like that feeling. Probably some of them were uneasy with him because he was gay, a fact that they had soon became aware of. They perhaps felt that the usual male bonding actions like back-slapping were inappropriate.

When I was CEO, I had managed the board by being very well organized for meetings, making sure that the board book was crammed with information. There had been some bullies on the board and I had felt hassled by them when I first became CEO, but once we got Visudyne approved and our stock price went through the roof, our board members who came from a pharma background kept silent on anything we recommended. These men all made several million dollars from their options just by sitting on our board and attending a three-hour meeting every three months. Now they had someone new to contend with. I got the feeling that many of the board members were biding their time before passing judgment on our new CEO. It seemed to me that they didn't warm to him.

Paul's style was open and very definite. He was assertive and had a tendency to think out loud, rather than thinking through an issue before he addressed it. I liked his style, because you always knew exactly where you stood with him and how he was thinking about issues. And no matter how definite he sounded about his opinions on a particular subject, I knew he would listen to arguments and that he would, on many occasions, be ready to change his mind. Logic always won the day with him. He expected people to challenge him.

In management meetings, Paul was equally open. He could be outspoken and sometimes could be inappropriate in what he said. He was a man without much of a filter. The friendships he forged with his management team were warm and almost too open. I realized that he trusted the management team and was treating them like they were family. He joked around with them as if they were old friends, when they weren't. I wondered if that much openness could bring him grief at some point.

But 2002 was the honeymoon period for Paul. Things were starting to move towards building the pipeline and growing the company.

CHAPTER TWENTY-NINE

I Thought These Problems Would Be Transient

In the spring of 2003, with the help of one of our investment bankers, we discovered Atrix, a company based in Fort Collins, Colorado. It was a company with a platform technology, a basic technology or mechanistic process that could be applied over a broad swath of product opportunities.

Their platform involved a way of formulating a variety of different drug products, and I was excited by the possible applications of that technology. They had developed polymers that would biodegrade in the body at controllable rates, depending on the tightness of the bonding between the molecules that went into forming the polymers. The active drug embedded within the polymer scaffold would be released as the formulating material biodegraded.

The company did very little discovery research. Their business model was to select already generic or soon-to-be generic drugs and reformulate them in their unique polymers to improve the quality of drug delivery. The newly formulated drug could be patented as a new chemical entity.

As an example, Atrix had a product, Eligard, in phase III trials for patients with prostate cancer. The active drug they were using was currently on the market, and due to go off-patent in a year. Men with this cancer are often treated with hormones that inhibit the production of testosterone, the hormone that stimulates the growth of prostate cancer cells. Successful inhibition of testosterone requires that constant levels of the inhibitory hormone be present at all times. Atrix had three different formulations of this product, one that lasted one month as well as ones that lasted three and six months. The product containing the inhibitory hormone was injected as a single intramuscular dose. The depot of the hormone in the body would biodegrade very slowly, releasing the required amount of hormone as the degradation took place. The shorter-timed products were recommended for men first going on treatment, so that effects could be assessed over the shorter time intervals before they went on to the longer-timed depot.

They had several other products in development, including an acne topical product that was also in phase III trials.

I was enthusiastic about Atrix because of what I saw as the potential of these polymers. In the future, they could be used to deliver a whole new array of biotechnology products that work best when there is a constant level in the body, many of which have very short half-lives in the body. When medications are given in pill form or injected as a bolus, the drug is absorbed in the gut, peaks at a given time in the blood, and then decreases as the drug is either digested or excreted. For antibiotics or other drugs, this fluctuation is less favorable than maintaining a given level over a period of time.

I also saw the possibilities of using these polymers as scaffolds for regenerating skin in severe burn cases or in other forms of regenerative medicine. The Atrix scientists already had some evidence that such possibilities were feasible. I was looking at the potential of where therapies might go rather than where they had been.

The Atrix CEO was a rangy, proudly Republican Texan with a booming voice to go with his big frame. His name was David Bethune. I had crossed paths with him many years earlier when he was a junior executive at American Cyanamid. But I hadn't known him well. He had now come to a stage in his life where he wanted to cash out and retire. He had a ranch and was looking forward to hunting and fishing. That was why he had the company up for sale.

From the beginning, I had concerns about corporate culture differences between Atrix and QLT. Bethune was running a pretty regimented company. I got the impression that he kept his senior management under tight, almost military control. His chief operating officer, Mike Duncan, was ex-military and conducted himself like a sergeant. Mike had the appearance of a tough guy, slightly overweight, his hair parted in the center and bristling out stiffly on either side of his head, making him look aggressive for some reason. But I liked him. He was a decent honest man with a soft side.

When we visited the company I was shocked but not really surprised to see a rifle and handguns in Bethune's office, in full view. But this was Colorado; there were probably guns in other people's offices too.

Not everyone at QLT was happy about Atrix. I had sided with Paul in favor of the acquisition. To me it represented an interesting platform and financial stability for us. Six-month Eligard promised to be a big product. I knew the marketing people were hoping we'd buy a company with a sales force and products they were marketing themselves. Atrix's furthest developed product, the prostate cancer drug, was partnered already, with separate marketing partners in Japan, Europe and North America, so there was no opportunity for developing a sales force.

But the merger went through and we became the owners of a Colorado-based company which we renamed QLT USA. The stock market reacted neutrally to our acquisition. They regarded Atrix as a specialty pharmaceutical company and didn't see what I felt was the exciting potential of the technology. A specialty pharmaceutical company is not given the same valuation as a biotechnology company by the stock market. Although the stock didn't suffer a big sell-off, it didn't shoot up either.

We embarked on the journey of incorporating the Atrix employees into the QLT structure. I saw for the first time some ugly qualities in some of my fellow employees. I suppose we are all subject to feeling threatened when we fear we might become redundant or even superseded. The companies had equivalent departments and each department was the responsibility of a director or vice president. Changes were inevitable. We were the acquirers and some of our senior and middle management felt it necessary to flex their muscles in ways that I found distasteful. After all, we hadn't been at war with Atrix. They were not the vanquished. But at all levels, I saw this vying for pecking order.

I know Paul was also annoyed by this competitive attitude. His view was inclusive, as was mine. We weren't looking to reduce staff.

We had no intention of closing the Fort Collins site. They had an operational manufacturing facility that produced the gel and the other delivery systems, which we did not. But there was conflict in most departments, with the exception of the preclinical research people, who seemed happy to interact and exchange scientific ideas.

There wasn't much redundancy between the Atrix and QLT research groups. The QLT scientists were doing a lot of discovery research using animal models and biological systems, whereas the Atrix people had expertise in polymer chemistry and formulation. We were short of formulation expertise in this field at QLT so the merger was very complementary. The scientists saw how they could benefit from each other.

The most troublesome department in terms of integration was clinical. Mohammad had built a fully structured clinical department that matched the clinical departments in big pharma. He had built it over the years and had statisticians, clinical writers, full quality assurance and quality control functions, and a large number of clinical research assistants at various levels of seniority. The department had swelled while we were completing the Visudyne clinical trials and writing up the material for filing. Whatever the problem, tensions were obvious between the QLT and Atrix clinical division from the beginning. I hoped these problems would be transient.

Mohammad was now spending most of his time on corporate activities in the company, so he left the handling of some of the difficult issues to Mo Wolin, who had not had a lot of experience in administration. Mohammad had rigorously trained a lot of the clinical people at QLT, and he had taught them the big pharma way of conducting clinical trials.

The Atrix people had had to be more creative in how they approached their trials. They had not had a lot of money and were consequently more efficient at making their dollars go a long way. David Bethune had been a demanding boss and had kept the company lean. Their chief of clinical was a dentist by training and a Mormon elder. I liked him; he was practical and non-judgmental. He knew how to deal with the FDA and had a good track record. Unfortunately, the two groups clashed continuously about small issues like how to prepare documents, and sitting in meetings with them became decidedly unpleasant.

While we were struggling with the issues around Atrix we got some very bad news about Tariquidar. The phase III trial had been moving along very well, we thought. I had been checking with Graeme Boniface, the clinical director in charge of the trial, about patient enrollment. He had told me that it was going fine, accruals were coming fast, particularly in Poland where the incidence of lung cancer is very high. But, he said, when he looked at the blinded data, he'd noticed that a number of the patients were dying soon after they entered the study. I knew these patients were pretty sick, so that didn't seem surprising, but then we got the bad news: there were more patients dying in the group that was receiving our drug. The data safety monitoring committee, which was privy to the unblinded data, saw that the patients receiving Tariquidar were dying from what looked like treatment-related toxicities.

We closed the trial while we reviewed the data. Then we discontinued it and gave the drug back to Xenova. The data indicated that patients on Tariquidar were dying as a result of the dose of chemotherapeutic drugs used. Tariquidar had been too effective. Because normal cells in many organs produce P-gp, they too were affected by Tariquidar and rendered more susceptible to chemotherapeutic drugs. Effective chemotherapy is a balancing act between killing cancer cells without killing normal cells, and clearly we did not have the appropriate drug doses to effect that differential killing.

I still question QLT's decision to abandon Tariquidar. I think we behaved like big pharma in dismissing a potentially useful product without looking at what the data were showing us. What the data said to me was that the drug was working too well. The flushing system of P-gp had been blocked so efficiently that normal cells were being rendered more susceptible to the chemotherapeutic drug, and deaths must have resulted from destruction of vitally important cells like kidney or gut cells that are high in P-gp. We could have avoided this outcome if we hadn't rushed into the phase III before we assessed the drug more carefully. Careful titration of Tariquidar or chemotherapeutic dosing would likely have uncovered the window in which benefit would be seen. By not exploring different doses in an additional phase II trial, we hadn't given the drug a chance. We could have gone back to the FDA with an amended protocol with lower starting doses of the chemotherapeutic doses.

But we didn't. We had succumbed to the pressures of impatient shareholders.

Today, in 2020, Tariquidar is still in investigational use and is in a number of studies sponsored by the NCI in the United States. It is generally agreed that the drug is a potent inhibitor of P-gp, but no one has yet completed studies that would justify its approval. Our returning the drug to Xenova created a taint around the product. This kind of failure is all too common in biotechnology, in big pharma as well as small companies. Big pharma always has a robust pipeline and they are usually ready to go on to the next product in development. With small biotech companies, a frequent cause for failure is not having sufficient funds to carry out all the early-stage trials as well as a big phase III trial, so they skimp and keep their fingers crossed. It is extremely difficult for a small company to mount new trials for a drug that has encountered a phase III setback.

We got the results from our first trial with Lemuteporfin for benign prostate hyperplasia (BPH), and they were exciting. The findings were very promising. First, they showed that patients tolerated the treatment extremely well, with virtually no discomfort reported following treatment. The physicians involved were also very happy with the outcome. Patients reported a durable response to the treatment with relief of symptoms and improved urine flow that lasted out to three months.

We decided to run a blinded study. I sat with the project team as they discussed possible protocols. The way that teams operate is that they meet as a group and first discuss various approaches to trial construction. There were many ways to design this trial. A comparative trial could compare our treatment with one of the standards of care – TUNA, TUMP, or TURP. These were the existing interventions for BPH, which involved means like heat or crude surgical procedures applied transurethrally to shrink the prostate. This kind of study could not be double-blinded, because the procedures were different and the treating physician would know which treatment the patient had received. Improvement in urine flow could be assessed more or less quantitatively, and the patient's evaluation of symptom improvement and level of discomfort was subjective but likely not biased towards either procedure. The end point for this study would be to show equivalent improvement in both arms of the study. All we would need to seek approval for our procedure would be to show it was not inferior to other accepted procedures. That should not be difficult to attain, since the other procedures were not that good. Also TUNA, TUMP, and TURP resulted in patient discomfort and prolonged catheterization.

Alternatively, a very elegant double-blinded study could be set up in which a placebo could be introduced into the prostate of control patients instead of our drug. The patient would be put through exactly the same procedure using saline, with the balloon catheter, injection, and light treatment. The end point in this study would be to show superiority of symptom improvement over the control.

The BPH team did not agree as to which way to proceed. The project manager and the scientific members all wanted to run the comparative trial against either TUNA or TUMP. They argued it would be much easier to show equivalence with one of the other therapies because these treatments were not very effective. Our treatment had an opportunity to be superior and would certainly be less uncomfortable than the alternative treatment.

The clinical people on the team preferred the double-blind design. They argued that this was an elegant method that would let us truly assess the value of the therapy. They argued that the FDA would surely insist on such a trial. While I liked the double-blind construct because it was scientific and would provide a real evaluation of the treatment, I realized how dangerous such a design was in the short term. The competing procedures for BPH had never had to go through a comparative trial, and we could find no references to comparative studies for any of the procedures. And they were being used clinically. Both TUMP and TUNA are procedures involving a device rather than drugs, and don't have to go through the rigorous testing that drugs do. Equivalence would be much easier to achieve than superiority. Our scientists pointed out that the placebo treatment with the balloon catheter and saline also constituted a kind of treatment. The balloon when inserted into the urethra was inflated prior to insertion of the syringe and drug or the fiber optic. Balloon catheters on their own can be used to expand blood vessel passages in patients with atherosclerosis. The act of inflation might cause temporary symptom relief for patients with BPH by expanding the urethral space. There could be an enormous placebo effect. It would be far easier to show that our treatment was equivalent to a procedure that we knew wasn't particularly effective, than to show we were statistically superior to a procedure that might well show some short-term benefit in a study known to have a big placebo effect.

Feelings became quite heated arguing this case. Clinical people felt they should have the final word because they represented clinical studies. Both sides dug in and finally resolution involved bringing in senior clinical people. They of course backed their own staff and we ended up going for the double-blinded study.

I remember sitting in Agnes Chan's office after the decision. She was the project manager for the BPH team. She had done her PhD with me some years earlier, and we were friends. She had started at QLT in molecular biology and had moved after a while into clinical, where she learned a lot about clinical trial construction. Finally she'd moved to project management. She is one of the most able people I know, with a perceptive analytical mind. I remember her shaking her head that day and saying that the clinical trial design that had been decided on, using a placebo as a comparator, was going to fail. I said I hoped she was wrong but I didn't feel too optimistic. We had created a serious problem for ourselves.

This year, 2003, had been a difficult one. I am by nature a very optimistic person, but I felt uneasy about the ongoing conflict between Atrix and QLT staff and deep worry about the BPH decision. I was reminded how easy it is to get things wrong in this business. And how rare it is to get everything right. But I had faith in Paul's ability. We had gotten everything right once. Surely we could do it again.

CHAPTER THIRTY

Ongoing Headaches

QLT had changed by 2004. It was no longer the focused research-based organization it had once been. Commercial expectations now dominated. We continued to look for in-licensing opportunities. I worried we were spreading ourselves too thin. We still had issues around the Atrix acquisition that had to be resolved. We had an urgent need to strengthen our ophthalmic franchise because we were soon to face serious competition.

We continued to experience difficulties in successfully merging two different corporate cultures. As a perpetual optimist, I had expected that people would accommodate differences and work out compromises, so that the company could work towards great success. I was to be proven wrong.

We knew we would soon face our first serious competition for our treatment for AMD. Two companies, Genentech and Eyetech, were in advanced clinical trials with anti-VEGF molecules. In the spring of 2004 Eyetech, the company that owned Macugen, released results for its phase III trials for wet AMD. The results were not as good as the results we had achieved with Visudyne. Patients continued to lose vision while they were on Macugen treatment, but they lost significantly less than placebo-treated patients. With Visudyne, patients' vision stabilized in the second year and most did not require further treatment. With Macugen, the treatment effect was lost after one year and patients continued to lose vision.

But the results were good enough for them to earn FDA approval, and they had a powerful marketing partner in Pfizer Pharmaceuticals, so we now had competition in the marketplace rather than on the horizon.

Andrew Strong, our senior director in charge of ophthalmic clinical research, who had been the early and highly effective champion of our AMD program, and Mohammad, his boss, started having disagreements. When I was CEO, I had urged Mohammad to promote Andrew to the position of vice president, making ophthalmology one division of clinical. I felt that he deserved this kind of recognition for having successfully steered Visudyne through phase III trials and for forging excellent links with the ophthalmology community. Our investigators all respected and liked Andrew enormously. He had shown quality leadership in handling the robust egos of his chief investigators.

Now that Visudyne was approved and ophthalmology was the medical area QLT was best recognized for, and we were actively trying to expand our franchise, Mohammad stepped in to take control of the program. He had resisted my suggestions about promoting Andrew. Even though I was CEO at the time, I didn't feel I should interfere with Mohammad's internal decisions. I urged him to find a way to promote Andrew in recognition of his skill and leadership, but I failed to convince him.

That year, Andrew and Mohammad had a deep disagreement about a protocol for a clinical trial we were going to undertake to expand Visudyne usage. Naturally, Andrew lost the argument. Mohammad was the boss. Andrew was a very thorough scientist and he had spent hours analyzing the data we had. I trusted his judgment, and when he told me he thought the protocol we had selected would not succeed I believed him. Andrew turned out to be right and our opportunities to enlarge our ophthalmic franchise were seriously weakened.

An opportunity arose in 2004 that Paul took advantage of. A local company called Kinetek was running out of money, with not much likelihood of refinancing. A friend of mine from the University of British Columbia, Steve Pelech, had started the company, intending it to be a service company producing test kits of reagents for identifying cell-signaling pathways. I had agreed to sit on their board of directors when Steve was CEO. The original business model was about producing these reagents for academic researchers who were studying the pathways. As the company developed, it brought in a new CEO and a new financial adviser, who changed Kinetek into a company that in-licensed a molecule that was clinic-ready. On the basis of this in-licensed product, they were able to raise enough money to carry on and also to continue research on the signaling pathway molecules and reagents.

Unfortunately for the company, they went through two bad CEOs and a failed trial. Their final CEO was an old friend of mine from Beaufour Ipsen days, André Archimbaud. Even André's business acumen could not save the company from running out of money.

Kinetek's technology platform went up for sale at a rock-bottom price. We saw that the company had several desirable assets. They had some very talented scientists, they had one very interesting platform, and they had amassed a significant asset in the form of research and development expenses. When companies spend money on R & D they keep track of those expenditures, so that when they start making money they can offset the R & D losses against taxable income. By buying the company, QLT stood to save tens of millions of dollars with the Kinetek tax credits. After buying the company, we ended up about ten million dollars ahead. And we had the interesting platform.

This platform concerned a molecule called integrin-linked Kinase (ILK). ILK is a protein that is often expressed in high concentrations on cancer cells and, based on animal studies, has been associated with enhancing tumor growth. Kinetek was therefore developing ILK-inhibitors as possible cancer therapeutics. These inhibitors had shown efficacy in a number of mouse and rat cancer models. We made the decision to continue with this research program at QLT, and some of our researchers began looking at ILK inhibitors for other conditions. We found that skin from patients with psoriasis contained enormous levels of ILK, so possibly an ILK inhibitor could be used to treat that condition.

We started another relationship in 2004. An ophthalmologist named David Saperstein phoned me. He was at the University of Washington and had been one of our investigators in our phase III Visudyne trials. He told me he had a collaboration with a scientist named Kryzystof Palczewski and they had started a company called Retinagenix. David said they had a drug for the treatment of a rare form of retinitis pigmentosa and that he'd be interested in talking to QLT about licensing the product.

I knew very little about this dreadful condition, but after his call I started reading up on it. Children born with this genetic defect start losing vision at an early age, and by the time they reach six or seven most of them are legally blind. Vision loss starts out at the periphery of the field of vision and progresses over time. The field of vision becomes smaller and smaller, causing classic tunnel vision. Usually before these patients reach maturity they have almost no vision.

David Saperstein was working with a form of retinitis pigmentosa called Leber's congenital amaurosis. The disease is caused by mutations in a series of enzymes that affect photoreceptors in the eye. The disorder hits the cones – the parts of the eye containing the photoreceptors (light-absorbing areas that permit light/dark distinction) that affect peripheral vision. A compound called 11-cis-retinal, produced from vitamin A by a series of enzyme reactions, is critical for the normal functioning of the photoreceptors. Patients with congenital amaurosis have a mutation in one or another part of the enzyme cascade that goes into regulating and producing of 11-cis-retinal. Lack of 11-cis-retinal in the cones causes degeneration of these ocular structures and ultimately the loss of their function.

David and his colleague Kris had developed a compound that, when it was administered, accumulated at the desired cellular target in the eyes of mice and could restore photoreceptor function. They had tested the compound extensively in a mouse model of the disease. Vision could be restored quickly in mice that carried the same mutation as human patients with this disorder, when the mice were dosed with the drug that Kris had synthesized.

Mo Wolin and I drove down to Seattle and met for most of a day with David and Kris. Kris was a brilliant scientist, and he had developed the drug and run the animal experiments. He was very excited about what he had accomplished after many years of research. I remember him saying, "You give patients with this disorder the drug in the morning and by lunch time, they'll be seeing." That was quite a compelling thought.

David had a large clinical practice and also taught at the university. Kris was a dedicated scientist and wasn't interested in business. The university owned the patents for this compound. Although they had formed a company, neither David nor Kris was interested in the process of drug development.

Being an immunologist, I had a long relationship with mice. I knew already that albino mice, lacking pigmentation, were essentially blind, and yet in cages they functioned very well and seemed to do as well as seeing, pigmented mice. I had been told that their whiskers acted as antennae for them, enabling them to "see." So I asked Kris how they could tell that they had restored vision to their congenitally blind mice.

The test was simple. Lab mice are usually kept in plastic boxes, much the shape of shoeboxes but larger. Standard cages have a metal grid top with built-in space for a water bottle and food container. If the top is removed, active mice will always climb to the lip at the top of the cage to explore the outside world. Blind mice will climb up but will not jump out of the topless cage unless there is an adjoining cage or some solid structure that they can sense with their whiskers. So in the vision test for Kris's mice, another cage is placed near the cage containing the treated mice, close enough for them to jump to, but not close enough for them to "see" with their whiskers. A seeing mouse will jump to the adjoining cage whereas a blind one will not. Mice are naturally very curious animals and always ready to explore. Kris's treated mice tended to make the leap.

I was enchanted with this opportunity. Mo also was intrigued and impressed. I knew that the disease was very rare and that any company developing a drug that treated it would not make millions of dollars from that product. But the thought of actually being able to restore the vision of people doomed to blindness from an early age was very compelling. I told David I would do what I could to persuade my colleagues to take on this opportunity.

Paul was also willing to go ahead and make a deal with Retinagenix. Our business development people started negotiating for rights. The deal proved quite difficult to complete, but we eventually did negotiate successfully. It turns out that there may be other less drastic ocular conditions that the drug can be used for. Night blindness and the difficulty that older people have in

light/dark adjustment are also conditions caused by a loss of efficiency in making 11-cis-retinol, and it is quite possible that the same
drug could be used to help older people with night vision in general, and night driving in particular.

Towards the end of 2004 it became clear that some of the difficulties experienced after the Atrix/QLT merger were not being resolved. The differences between our clinical/regulatory staff and
their Atrix counterparts continued to fester. All the other departments seemed to be operating smoothly. But the big pharma culture
that had been deeply inculcated into the QLT staff was hardening
rather than abating.

Atrix also continued to claim ownership of their own programs
and rebuffed efforts from QLT to interfere and take control. The
Atrix clinical department was a very lean organization, akin to what
one might expect to find in a biotech company much smaller than
QLT. They designed clinical trials that were perfectly adequate and
not sloppy in any way, but they always rigorously controlled costs
and trimmed all unnecessary extras out of protocols. I liked the approach they took.

I found it really unpleasant to sit through some of the clinical/
regulatory meetings because of the conflict between the teams. I observed an unexpressed feeling of superiority coming from some of
the QLT people, countered by mute resentment from the Atrix people. Either Mo was unaware of these undercurrents of ill will, or he
chose not to rein some of the people in. I spoke to him about what
I had observed and he seemed to be taken by surprise. He hadn't
noticed anything inappropriate. I don't think he ever took my comments seriously enough to try to remedy the situation.

On the other hand, Bob Butchofsky, who had become the vice
president of marketing after Lee Anne's departure, had created
harmony. I had been impressed by the quiet efficiency with which
Bob had managed the Ciba Vision relationship. Now, in his role as
vice president, I saw how much more cohesive and happy the people in marketing had become. He had a very pleasant manner and
was non-directive, and had no obvious difficulties with salespeople
from Atrix. But he could be tough and stubborn when necessary.

He wanted to see the ophthalmic franchise at QLT grow, and
brought a possible opportunity to the company in the form of an

ocular product for treating glaucoma from a small start-up he knew in San Francisco. He was very excited about it and tried hard to interest management as well.

Glaucoma is a disease of older people and may cause serious vision loss if it is not treated. It is caused by failure of the eye to successfully clear a lubricating fluid that circulates through a chamber at the front of the eye and is cleared through a fine meshwork of tissue around the edge of the iris. As the eye ages, the clearing mechanism may slow down, causing a buildup of fluid in the eye and an increase in intra-ocular pressure. The increased pressure may cause damage to the optic nerve.

There are many treatments for glaucoma, usually in the form of eye drops containing drugs that either enhance clearance or reduce intra-ocular pressure by other means. Because the disease is basically without painful symptoms, people frequently discontinue use of eye drops because they think they're already better. Non-compliance with treatment on the part of patients with glaucoma is a significant problem for ophthalmologists. Bob's friends in San Francisco had created a tiny container called a punctual plug that could be inserted painlessly into the corner of the eye adjacent to the nose. The device would be filled with the drug of choice, which would then release the treatment over a prolonged period of time into the eye.

He was very enthusiastic about the plugs. Bob was a marketer and was certain that these plugs could become a multi-million dollar product. He had done the research. He wanted to see the company focus more closely on ophthalmic products. I could understand his point of view. QLT was beginning to look to me as though we had no focus.

While 2004 was proving to be a difficult year for the company, things were happening in our private lives that brought Ed and me great joy. Jennifer and her husband Paul had become concerned about there being too many people in the world, so they decided they would adopt a child, and then decide whether to have a child biologically. They adopted a child from India. They researched agencies and went through interviews and mountains of paperwork, and finally were blessed with a baby girl. They went to India to pick her up in February. Ed and I went to New York the week Nikita arrived. I wanted to see this child while her past in the orphanage still dominated her behavior. I wanted to see what she could tell me.

Nikita was fifteen months old, and weighed just sixteen pounds. She was not walking but getting around by a kind of crawling. Her little legs were rubbery and she clearly had not been encouraged to walk or even stand. I fell in love with her. She was so grave and her look was so penetrating. Apparently she had been very upset when Jennifer and Paul took her away from the orphanage and wouldn't eat or drink anything the first night they had her. They had been very distraught and contacted us from India for advice. I told them that I thought Nikita's sadness was a very good sign that she had formed strong emotional ties with her caregivers at the orphanage. She had clearly been well cared for at the orphanage, even if they had not encouraged her to walk. I also liked the fact that she had a strong will.

Nikita was very set in her ways. She went to sleep at 7 p.m. and slept for twelve hours. She talked in her sleep and even shouted sometimes. She never cried. And at the beginning, she hardly ever smiled. She was terrified of the bathtub and screamed when we tried to encourage her to get in. This phobia disappeared quite soon. I suppose she'd never seen a bathtub before she arrived. She didn't know how to drink from a bottle, a disturbing problem for Jennifer, who was trying to get as many calories into her as possible.

I tried to see the world through her eyes. How strange it must seem to her, to come to this cold place and be surrounded by these pale people speaking words she had never heard, and certainly had no comprehension of. No wonder she spent her time looking thoughtful. But she was beautiful, with her raven-glossy curls, her big eyes, and dark chocolate velvet skin.

That summer, our son Ben got married again. His children were now seventeen, fourteen, and twelve. His new wife, Deborah, had two daughters from a previous marriage, who were seventeen and twenty-one. Deborah was the principal of the Christian school in Gibson's Landing. Ben had changed his name to Levy and been baptized in Deborah's church just before the marriage. Jennifer and Paul came out for the wedding with Nikita, who was nineteen months old and now walking, smiling, and quite sure of herself in her small self-contained way. She enchanted the whole wedding party and had Ben's youngest daughter, Emma, and two of her friends trailing her through Ben and Deborah's spacious garden as she toddled around investigating. Ben's former wife and her sisters attended the wedding, as did Deborah's former husband. Deborah's elder daughter Leah stood up with Deborah and gave her away.

Just after they picked Nikita up in India, Jennifer and Paul decided that they wanted to have a biological child too. Jennifer became pregnant immediately. She was five months pregnant when we'd seen her at Ben's wedding. They had a little boy at the end of November that same year, so they added two children to their family within ten months. They named him Jesse. We went to New York in December to get acquainted with him. He was a beautiful baby who looked very much like his mother. Nikita, who was now nearly two, still wasn't talking, although it was clear that she understood most of what was said. Although she had been prepared for the arrival of the baby and had been trying to feed it through her mother's navel, she did not seem too happy with the reality of her brother's arrival. (Nikita and Jesse are now extremely close, but Nikita was once overheard boasting to her brother that she was adopted while he had had to be cut from their mother's belly.)

It was wonderful to be able to get away from the troubling issues around QLT for even a short while and to spend time with Ben, Jennifer, and Paul and share their joy in building their families. I had a great deal of confidence in Paul Hastings's abilities and his positive attitude as the new CEO, but he'd need cooperation from his senior management, and I was no longer sure that he could count on it.

CHAPTER THIRTY-ONE

He Felt Thoroughly Betrayed

By 2005, QLT had accumulated enough potential product assets to keep us busy for quite a while. Our next-generation photosensitizer, Lemuteporfin, had been selected to go into formulation for topical applications. We had some strong preclinical data that suggested it could be used to treat severe acne. The six-month data from the Eligard prostate depot drug was due to be analyzed and submitted for approval. The FDA had already approved the one- and three-month depots and it was anticipated that the six-month data would be approved too. The marketing partners were not going to launch the product until the six-month dosing was approved, since that was the product with the greatest appeal. There was already a three-month product on the market.

The results from the Atrix topical cream for mild to moderate acne were analyzed and shown to be significant over the placebo, so that product too was soon to be on the market. It had been partnered, so we would receive a royalty stream for it.

The marketing people at QLT were not happy about having nothing to form a sales team around, since all the close-to-market products were already licensed out to marketing partners. Some of them had been against the Atrix merger from the beginning. Bob Butchovsky, who had been unsuccessful in convincing others at QLT to in-license the ocular punctal plugs, told me that his friends in San Francisco who had brought him the punctal plug opportunity were now thinking of forming a company around the technology, and wanted Bob to become its CEO. He was torn, he told me. His kids were growing up Canadians and moving into high school. He liked our company and the people in it, but he was excited about the opportunity to be a CEO. He was at that stage in his career where he was ready to make a move and feared that if he delayed, he'd be less inclined to do so. I could understand his position, and the appeal of starting a company on his own. He had found a technology he was excited by, even though he hadn't been able to arouse the interest of anyone else at QLT.

I could see that my role in the company was changing. I began spending more time with the research people and less with management. The disunity within management troubled me. I think the reason I migrated towards the research group was that I felt alienated from management. I had strongly supported Paul in his decision to merge with Atrix, whereas others had been less enthusiastic. Management meetings struck me as fraught with unstated grievances. The apparent lack of openness made me feel more and more uncomfortable. At the same time, my interest in participating in the running of a commercial company, which was never great, was flagging further. I found scientific discussions of potential applications of our products far more interesting. I thought the core technology from Atrix, slow-release formulations, had a huge potential to formulate the products coming from molecular biology – biological molecules that clear the body fast and need a maintenance level to be therapeutic. I began going into the office only four days a week. I was seventy now, and although I didn't feel tired, I wanted more time to myself. Having a long weekend every weekend felt good.

Ed had retired from QLT, and spending the extra time with him was important. In addition to participating in research projects at UBC he wanted to become more involved with researching charitable giving, and running the family foundation we'd formed a couple of years earlier. We were getting to know the network of people in Vancouver with the same instincts about philanthropy that we had. That, too, was a pleasant diversion from the business world. Certainly, we found we had more in common with these people than we did with many of our business associates.

Research at QLT on ILK inhibitors was leading to a number of exciting findings. While ILK inhibitors had originally been developed with the goal of finding a cancer treatment, we were not impressed with the anti-cancer effects of most of the anti-ILK drug candidates we tested. We were having difficulty repeating some of the data generated by our academic collaborators. On the other hand, we were getting really interesting results for other possible applications. We knew that the skin of psoriatic patients had excessive levels of ILK in comparison to normal skin. A topical ILK inhibitor might be effective in treating that condition. Early work in the area of anti-inflammatory properties of ILK inhibitors looked promising, increasing the likelihood of its being effective in psoriasis.

Then Genentech made an announcement about their phase III data for Lucentis, their drug for treating macular degeneration. Their results were spectacularly good. Patients on the treatment arm of the study had, at the end of one year, an overall average improvement in visual acuity. There was no question; their results were better than ours had been. The only negative aspect to their treatment was that the anti-VEGF agent did not appear to arrest the condition and patients continued to require treatment every six weeks. Contrary to our initial thinking, patients tolerated injections into their eyes at this frequency.

Genentech would clearly get FDA approval for Lucentis sometime in 2006, and without doubt that would seriously erode Visudyne sales. This threat did not improve the atmosphere at QLT. We now had ample reason to regret that we hadn't gone out on our own, independent of Novartis, and licensed an anti-VEGF reagent to carry out a combination trial of anti-VEGF plus PDT. I was convinced then, as I am now, that this combination would work better than either therapy alone and save Medicare millions of dollars. I have spoken over the years with individual retinal specialists who have worked out protocols with combination therapies and they claim that after one or two treatments with the combination, patients require no further treatments. But a clinical trial was never carried out, so combination therapy has never been adopted as a pattern of practice.

It was customary for QLT to hold their summer board meetings at a resort, so that board members could socialize and play golf. The two-day event was structured to encourage open and informal discussion between board members and senior management during the scheduled free afternoon. Partners of board members were usually invited.

This year the meeting was held at Whistler Mountain, where any number of activities was available – golf for the staid elders and ziplining for the more adventurous. And there was always the spa. On the free afternoon I refused invitations to accompany others and went by myself to the spa for a massage and some solitude. I can only describe the feelings around the meeting as edgy and full of tension. I got the feeling there was disunity amongst QLT management.

Bob spoke with me again about the punctal plugs opportunity during those two days at Whistler. He told me that his associates in San Francisco who held the rights were putting pressure on him. If Bob was unable to convince QLT to license the product, they wanted him to join them as CEO in forming a company around the technology. Bob said he was feeling a lot of pressure, and had talked this over with Paul after he had been unable to persuade enough people at QLT to pursue the punctal plug opportunity internally. Paul had been supportive of Bob, saying he understood Bob's position, understood his uncertainty. Now he was asking my advice. I thought Bob had some of the right qualities to be a CEO. He knew how to run a commercial company. He was a quiet, honorable man who led by example rather than direction. His managing of the marketing people at Ciba Vision had been tactful and efficient. He was in his mid-forties, an appropriate time to make such a move. I encouraged him to follow his inclinations. Bob is a thoughtful man and I could tell he was torn.

When the meeting was over and we were back at work, the edginess and tension among management continued. I knew something must be going on and suspected there was a split in the management team. Some people knew what was happening and others did not. I also sensed that, whatever it was, Paul probably didn't know. I knew that Paul's style had upset some of our management. He was enthusiastic about the Atrix acquisition and made significant efforts to be inclusive with Atrix management. I knew that battle lines were being drawn up by QLT staff to counter any efforts made by Atrix staff to be in charge of any ongoing projects.

Several days passed. Then one morning, just as I got to work, my assistant, Susan, told me Duff Scott, our board chairman, was on the phone. I wondered why he was calling me rather than Paul. We had a management meeting scheduled, so I told Duff I didn't have more than a minute.

Duff told me that he had received Paul's resignation as CEO earlier that morning. Jack Wood, our only board member living in Vancouver other than myself, had met with Paul at home earlier and would be attending the management meeting. He would link me in for further discussion regarding the resignation.

Instead of a management meeting, that morning we gathered to hear about Paul's resignation and plans going forward. Jack Wood had arrived at QLT while I was talking to Duff. Jack chaired the meeting and Duff joined by phone.

I couldn't believe what I was hearing from Duff. I walked into our boardroom in complete shock. I looked around the room at our senior management. I knew these people well and could read the surprise and concern on many of their faces. Linda looked away when I caught her eye, as did Therese Hayes, who had replaced Elayne in Investor Relations. Bob looked confused. Mohammad looked inscrutable. Mike Smith and Bill Newell, people that Paul had brought into business development after Ed retired, looked puzzled, as did Larry Mandt and Alain, our VP of project management.

Duff said the board would be meeting later that day to discuss appointing an acting CEO from the management team. Nothing was said about the reasons for Paul's departure.

I sent Paul an e-mail telling him how shocked I was about what had happened. We spoke briefly. He was also very disturbed about the way things had transpired. I realized that he was probably in shock, like me. We arranged to speak further later in the day.

Then Bob came to my office. He looked shaken. I suppose that all of us who had been taken by surprise were shaken. He had known nothing about what had been going on. He'd spoken with Paul the previous week and discussed his possible departure from QLT. He told me he'd finally made a decision and was planning on handing Paul his resignation that day. He had the letter in his pocket. He felt bad because others at QLT had known he was thinking of leaving, and he feared that his position had added fuel to the criticisms of Paul. He said he had no criticisms of him. Then he looked at me and asked if the choosing of the acting CEO was already decided. I said that as far as I was concerned, it was not. He asked if it was too late to throw his hat into the ring. I said I would be happy to forward his name and provide him with a recommendation. He said, "Then I'll tear up my letter of resignation."

Jack Wood, who was still at QLT, came to see me later. He was apologetic. He explained that Duff, as chairman of the board, had convened the board members without informing either Paul or myself. I was not included because I was still an employee of QLT, he said. The board (minus Paul and me) had been called to discuss what had been presented to them as insoluble differences between management and the CEO.

When I got in touch with Paul later, he said he felt betrayed by some of his colleagues. He knew that he had contributed to the general discontent of management towards the merger by his insistence on shared authority between Atrix and QLT managers. He had thought things could be worked out, that things would calm down. But I think Paul might not have realized how deep the resentment went. These ongoing issues had not only created a management problem, but also had given rise to a lot of complaining. Complaints landed up on the desk of our human resources vice president, Linda Lupini. Linda was both well-liked and respected at QLT. She had an open-door policy and people confided in her, so she no doubt knew more than most of us about perceived grievances on the part of management.

I wished there had been less destructive ways of solving the issues facing the company.

The board met by phone to discuss succession. Duff led the discussion and put forth Mohammad's name as the obvious candidate for CEO. I put forward Bob's name. When it came to a vote, a majority of the board sided with me.

By the end of the following week, Paul had received an attractive offer to become CEO of Oncomed, a thriving biotech company in the Bay area that was developing monoclonal antibodies to cancer stem cells. We discussed the technology and he realized how exciting it was. The company was well-funded and the technology was sound. He decided to take the position. I suggested he think about it. I was sure he'd get other offers. But he wanted to get the bad taste of what had just happened out of his mouth as soon as possible. Paul has been happy at Oncomed and has managed the company very well.

Duff came to QLT to make the announcement that Bob was to be the acting CEO. I was sitting on the podium with Duff when he spoke. Mohammad's face was like stone. I knew he would leave the company as soon as he found another position. Most of the employees were confused but not upset, because Bob was well liked.

I worried about how our discovery scientists would fare under the new management. While neither Bob nor Paul had a science background, Paul had a natural high interest in the research activities at QLT. I had not seen the same interest coming from Bob. Bob was a man who had spent his adult career dealing with the commercialization of ophthalmic products, and it was clear to me that his intent would be to strengthen the ophthalmic franchise we already had. I wondered about the fate of the Atrix merger. I didn't have a lot of hope for its success now.

Well, I thought, Bob will get his punctal plugs at least.

CHAPTER THIRTY-TWO

Problems with Activist Shareholders

By 2006, the only reason I had to stay connected to QLT was that I was concerned about the people in discovery research. I realized I had to distance myself from management, and although Bob encouraged me to remain involved I knew he and his staff had to develop their own pathway. I was more than willing to help in an advisory capacity, but only when asked. I guessed that Bob's plans for the company were to be more focused on ophthalmology. It was the area he knew and loved, and I knew we were spread too thin. Punctal plugs were back in the QLT portfolio. The deal with Retinagenix was signed, which made me happy. David Saperstein, the ophthalmologist from Seattle who had contacted me about the drug for Leiber's congenital amaurosis, was appointed as our chief medical officer. Andrew was made a vice president, a long overdue and deserved promotion.

But much of the discovery research in the company was not focused on ophthalmology, and I worried about the future of the non-ophthalmic programs.

I moved my office down to the research labs and familiarized myself more fully with the research activities that were underway. I got a greater appreciation of the quality of our research staff and was very impressed. Some of the programs were incredibly promising both scientifically and in terms of business opportunities. But they needed resources to continue developing, and I sensed that discovery activities not focused in ophthalmology were not high on the company's agenda. I doubted that any of them would be put into development unless they were partnered with another company.

Mohammad did not stay long at QLT. He found a position as CEO of a small California-based oncology company about six months after Bob's appointment as CEO, and I wished him well. He lasted a year in that company. Paul, on the other hand, was thriving at Oncomed. The company had raised a lot of money and the research was going well. I stayed in touch and saw him occasionally whenever I got to San Francisco or he made a trip to Vancouver.

It was in 2006 that QLT started running into problems with activist shareholders, as had another profitable Vancouver-based biotech company, Angiotech.

I understand the motivation and actions of activist shareholders in other businesses, usually long-established tired old companies that are profligate with profits and stagnated. Under this kind of scenario, hedge funds or other well-funded financial groups can buy up shares on the market until they have a sufficient number to literally hijack a board of directors, usually when the board slate comes up for re-election at an annual general meeting (AGM). In the absence of activist shareholders, AGMs are usually orderly affairs, with attending shareholders voting almost automatically to elect a recommended slate to the board of directors. The list usually includes most of the existing board members, with an orderly turnover of a few old members retiring and the appointment of new members preselected by the company or board. Most shareholders in public companies do not attend AGMs. They usually vote by mail or sign proxies for voting over to the existing board. Many shareholders don't bother to vote. If activists take over, they work to get shareholder proxies and take over at AGMs to elect their own board, giving them power to hire and fire within the company, which they do. They appoint their own CEO, sometimes just a puppet to keep things going until the business can be sold. These tactics have been used, sometimes to good effect, with aging companies that need a shake-up. Some activists will simply strip the company assets and allow it to wither.

The usual tactic after such a takeover is to look for redundancies and trim the company so that expenses go down and earnings per share go up, to the benefit of shareholders. Sometimes such a process can be beneficial to the company as a whole by breathing new life into it.

I had gone through one activist shareholder event with Anormed. The board was hijacked exactly as I have described here. I was on the board at that time and experienced the situation first-hand. The issue that generated shareholder activism with Anormed was that the company was almost finished with a phase III trial that looked promising. Their phase II data had already shown that they had a drug that worked. If they were to complete the phase III study and proceed to market their own product, they would have to raise more money, which meant dilution. In spite of the company's need to raise cash, it was an attractive opportunity to companies wanting to build their pipeline near term. If Anormed was for sale, they were likely to receive a substantial offer. QLT had wanted to merge with them a few years earlier, but they had wanted to go it alone.

A group of New York shareholders wanted to see the company sold. They had invested heavily in Anormed's previous round of financing. Anormed's employees and existing board wanted to see the company thrive on its own. The activists were able to convince a number of other investment firms who held large blocks of Anormed shares that selling the company was to the investors' benefit. The activists took over at the AGM and fired the board and the CEO. The company was sold, the technology and assets moved to California, and everyone at Anormed was out of a job. The acquiring company essentially junked Anormed's ongoing research programs. There was enough of a short-term benefit to shareholders that the sale was generally regarded as a great success. But if they had waited for the phase III results, the success would have been much greater. Anormed's departure was a significant longer-term loss to the biotech community in Vancouver, along with the scientific capital that was buried in the process.

The QLT activists were led by a couple of New York–based hedge funds and one in Montreal. By 2006, they owned a significant percentage of the company. Although I was removed from any interactions between Bob and the activists, I understood that they were putting a lot of pressure on him to downsize the company and sell off some of its assets. Bob went along with their initial demands and started with a reduction-in-force, streamlining departments by giving notice to a significant number of employees.

Reductions-in-force (RIFs) are a completely demoralizing activity in any company. I knew that QLT was overstaffed. Some departments were bloated and could afford to be trimmed. Also, taking these moves early on probably suited Bob's own plans to streamline the company. At the start of an RIF, staff is warned that one is about to take place and employees are first given a choice. People who are thinking of leaving are encouraged to give notice and are offered the standard severance package, which at QLT was generous. After the voluntary departures, if the required reduction in staff has not been reached, people are laid off. It is an agonizing time.

After the first RIF came demands that QLT sell off some of its assets. In the next year or so, QLT found a buyer for the Atrix manufacturing facility and the Atrix products that were now on the market and selling well. A South American company took over the operations. I wondered when the activist shareholders would have had enough. Then pressure was put on Bob to buy back shares in the company. When he followed the directives of the activists, they put further pressure on him to reduce staff again.

It was pretty clear to me that these activist shareholders would never be satisfied. I didn't envy Bob's job. I didn't understand their end game. I still don't.

Then the three-month data from QLT's benign prostate hyperplasia trial came out. As Agnes had predicted, our treatment did not show a significant benefit over the placebo, although there was a trend towards significance. Both arms of the trial showed significant improvement from pretreatment symptoms. The placebo effect had been impressive, just as Agnes had predicted. There was an option in the protocol to extend the observation period out to six months. Bob asked my opinion as to whether we should continue the trial for another three months. I said I thought the placebo effect would likely diminish faster than the treatment effect, so it was likely we'd get significance at six months. He didn't follow my advice, and closed the study, further diminishing the company's research activities.

The only bright spot for me during those years was the retinitis pigmentosa therapeutic that we had licensed from David Saperstein. The drug passed a significant milestone. There is a breed of dogs that have the same mutation in their photoreceptors that patients with

Lieber's congenital amaurosis have. These animals are considered the critical test that drugs for this condition have to pass. The QLT drug was used to treat puppies. Their visual acuity was assessed after treatment. The miracle happened, and the puppies regained their sight. This made news in the ophthalmology community. I was happy for David Saperstein.

Similar to mice, blind dogs do quite well getting around by using their noses and whiskers and by knowing their environment. Many elderly companion dogs lose their visual acuity as a result of cataracts or diabetes, and often the owners don't realize that their pet doesn't see until they are put into unfamiliar surroundings. As long as their environment is stable, they get around by whiskers and nose.

The way vision is tested in these particular puppies is that they are acclimated to the test space by letting them play together in a room in which boxes are placed in certain locations. As the puppies familiarize themselves to the room they become accustomed to the places where the boxes are placed, and automatically avoid them when they move around. After treating the puppies with the Retinagenix drug, they were allowed back into the test area, although the boxes had been moved. Unsighted puppies will automatically move around the room avoiding where they have learned the boxes are located, and will bump into the newly placed boxes. Sighted puppies will avoid the newly located boxes because they can see them.

These results were thrilling. I was very happy we'd put the effort into getting the drug ready for clinical testing.

Then the moment that I had foreseen but dreaded arrived. Bob, under pressure from the hedge funds, decided to eliminate the discovery department of QLT, with the exception of the ophthalmic research group. All the remainder of the scientific staff was terminated. I quit the company that day and resigned from the board. There was nothing for me at QLT anymore.

It was about a year after that event that the activist shareholders took over the board, fired all the existing board members, and got rid of Bob. There was further downsizing, until there was only a skeleton staff left at QLT. One of the hedge fund CEOs took over as CEO of QLT. Interestingly, this new management hung on to the Lieber's congenital amaurosis drug. The product has now been tested in humans and has been shown to be active.

One person who stayed on at QLT was Susan Hall, who had been my executive assistant when I was CEO and served in the same role with Paul and then Bob. She and I had stayed in touch, so I had general updates on how things were going, although Susan was always the consummate professional. Susan had a very high salary at QLT and knew she'd have trouble matching it in any new job. But this part of my career was over

I was very sad to see the enterprise that so many of us had built from scratch with love and passion be reduced to a wraith and overseen by people who had no appreciation for or interest in the talented people and special skills that we had assembled. I was sad for the community in which I live and sad for the people directly involved. Unfortunately, the QLT story is far from unique. Publicly traded companies like QLT are industrial Don Giovannis. They sell their souls. In order to raise the huge amounts of money necessary to bring innovative products to the market, a biotech company has little option but to go public. Public companies are owned by their shareholders, and the shareholders' goals are not usually those of science-based companies. One is obliged to feed the beast. The first victim of the demands on a public company is almost always the research. Then when the company becomes profitable, the focus becomes earnings per share and, once again, research is caught in the crosshairs. To me, the paradigm makes no sense.

CHAPTER THIRTY-THREE

New Opportunity

I am not someone who wallows in regrets about past mistakes. I learn from them and continue my journey, hopefully to a more enlightened position. Leaving QLT when it was in this depressing decline was liberating for me. I would not have left if there had been any role for me there, but shutting down the research department eliminated any possibility that I could help the company. The time leading up to my departure had been very depressing. I felt very much for the people who were losing their jobs, many of whom were friends to me, and I did what I could by providing references and letters of recommendation.

After Ed and I retired, we made ourselves available as advisers and board members at local start-up companies, lending our names and knowledge to people who were trying to find how to convert scientific discoveries into important treatments. We invested in many small start-ups and continue to do so. I remain wedded to and excited by the scientific findings that create the vision of a new therapy.

One such opportunity came from QLT in 2008, and it resulted in the formation of a new company. I often met for lunch with a number of the scientists from QLT, especially those who had been graduate students in my lab. On this occasion, I met with three former colleagues and we talked about the very promising research program that Dave Hunt had led at QLT. Dave ran the dermatology research programs at the company and had come up with a number of very exciting leads. We spoke of how sad it was that none of these would ever be developed.

Then, a couple of weeks later, I had lunch with Dan Wattier, who had been the vice president of marketing at QLT after Bob became CEO. He had left the company about the same time that I did, and was looking around for something to do. During that lunch Dan and I came up with the idea of starting a company around some of Dave Hunt's interesting and novel dermatology products, which were now languishing and dormant as an undervalued asset. There were three separate lines of research in this portfolio. Each one could lead to at least one product. QLT was certainly not interested in developing these assets, so we believed we could get a license to develop them.

I could see Dan was excited by the possibility and we decided to take the idea forward. I said I'd be happy to be the new company's board chair and help them get financing if they formed a company by bringing Dan and the three scientists – Agnes Chan, Dave Hunt, and Jing-Song Tao – together. All three of them had been graduate students in my lab at the university. Each of them brought special skills to the new company. After leaving my lab, Dave Hunt had run the dermatology research at QLT and had discovered a unique molecule that had great promise as a possible treatment for severe acne. He had also developed and tested one of the ILK inhibitors that showed strong anti-inflammatory activity. Agnes had moved from research into the clinical department, and from there she'd become a project manager, so she was very familiar with all stages of drug development. Jing-Song was experienced with Lemuteporfin, QLT's next-generation photosensitizer. This was in May of 2008, and Bob Butchofsky was still the CEO of QLT.

We had to make a deal with QLT to license these products. We thought that would be easy. Dan and Bob were friends; surely, we could get the products as a spin-off out of QLT and QLT would take shares in the new entity instead of expecting us to pay for the license. QLT had no interest in developing these products itself. This exchange of license for equity in spin-off companies is the norm for these kinds of transactions.

We were wrong. Bob had appointed Alex Lussow as head of business development, which was responsible for out-licensing. I discovered that Bob's non-directive approach extended to a completely hands-off style with his senior staff. Alex was bent on getting up-front cash for the assets. He apparently valued that business model. We didn't know the reason, but these demands made it almost impossible for us to raise venture capital. Venture capitalists don't like to see their investments go right out the front door of the company to pay for rights to the products. They want to see their money invested in the development of the product. QLT was asking for an initial payment of a million dollars for the rights to the products. There would be no equity or royalties involved in the deal. We were very surprised and dismayed by his response. We hadn't expected to be asked to pay anything up front for the products. QLT certainly didn't need any money. If we had access to that kind of money we would have gladly paid it. But we didn't. What little money we'd been able to raise we needed to further the development of the products.

We all invested our own money in the new company. As well, we were able to raise an "angel" round of financing from friends and family. This small amount of seed funding (about half a million dollars) enabled us to go on the road and tell our story to venture capitalists in San Francisco and Los Angeles while we continued to negotiate with QLT. There was interest, because the products were novel and compelling and Lemuteporphin was clinic-ready. Novel products in dermatology are rare. We had several offers of financing if we moved to the United States and accepted an American CEO, but we didn't want to do that except as a last resort. I suppose the venture capitalists had reason to be dubious about Dan's credentials. He had no track record as a CEO and the rest of the team had no business experience whatsoever. While my presence may have helped, it was not sufficient to get these firms to part with their money.

We went quite a long way trying to merge with Dusa, a PDT company that was marketing a photosensitizer, Levulan (amino levulinic acid), for a PDT treatment of skin cancer. They had revenue but no pipeline. We had a pipeline but no revenue. I thought that, put together, the combined company would be able to raise money easily. After months of meetings and negotiations their CEO got cold feet and pulled out.

One group we met with controlled a fund dedicated to novel dermatology products. They were located in Los Angeles and were very elegant. The women were young, beautiful, and groomed to perfection. The men looked like models. We felt grubby in their presence. But they got very interested in our products. However, they would only invest under conditions in which they took control of the assets without including us in the new company.

We still had a lot of friends at QLT, friends who wanted to see us succeed. They formed our unofficial grapevine. We heard via this grapevine that Alex was being wooed by another possible buyer for the QLT dermatology products that we wanted. The wooers were none other than our glamorous friends from Los Angeles. They were trying to do an end-run on us for a license to the products we were trying to extract from QLT. I called Jack Wood, who was still on the QLT board, to ask his advice and to see if some members of the board might urge Alex to do business with us rather than the California group. Jack then asked me a peculiar question. He asked why we were reluctant to offer QLT share equity in our new company for rights to the products. I explained that we had hoped to do precisely that, but we had met a wall of resistance from QLT. There was a silence from Jack. Then he said he would take the matter up with the board.

Within a couple of days we heard from QLT that our original offer to QLT had been accepted. We celebrated. So, now we had our assets but still had almost no money.

We thought that perhaps if we found some high-profile American dermatology person to sit on our board we might be able to get the financing we needed to get the products into clinical testing. A man called Tom Wiggins came to mind. Tom was considered one of the gurus of the field. I've heard him described as the Steve Jobs of dermatology. He had managed to turn unattractive dermatology companies into attractive ones by combining them. More than once he had then sold the company, and in the process had made shareholders a lot of money. The companies he had created continue to flourish. Dan and I called him and had a brief conversation. I could tell he was interested. He said he was getting a group together to form a new dermatology company. He invited us to meet with him and his associates in New York and to give a presentation.

Dan, Agnes, and Dave flew out and came back very excited. They had clearly been a hit. What Tom proposed was not quite the deal we had wanted, but it made such good business sense that we went ahead with it. He wanted our products to constitute the core pipeline of a new dermatology company he was in the process of creating. He liked the novelty of our drug candidates. Agnes, Dave, and Dan would have senior positions in the new entity but the operations would be in San Francisco, not Vancouver. And Tom would be the CEO.

We all had founder's shares and we had all invested money into our early financings. When the deal was completed, the founders and original shareholders got almost twice their investment back plus a lot of shares in Tom's new company.

Tom Wiggins had no difficulty raising the millions needed to get the products into the clinic. He had the track record and a brilliant management style. Even though I would have liked us to keep the products and do the development in Vancouver, I can't deny that this agreement was absolutely the best for getting drug development underway. The company, Dermira, did an IPO in 2014, which went very well. Now Dermira, which also licensed-in other products, is valued at well over $500M.

This is just one of the early-stage companies I have worked with. I still have so much fun with the bridge between beautiful scientific discovery and early data and its translation into something directly beneficial. Looking at data still excites me. I like giving my time and sharing my experience with young scientists who have that spark and desire to use their scientific knowledge to further our understanding of disease so that we can find methods for cure or maintenance.

Within the last year or so I have noticed many women in the Vancouver area who are forming biotech companies and becoming scientist/CEOs. I hope I had something to do with that.

Outside of our biotechnology activities, Ed and I formed a family foundation, ILLAHIE, some years ago. There are so many worthy causes in our complicated world that it is difficult to make choices, but we try to support actions that can effect systemic change. Most of our family is involved in the foundation. The exception is Jennifer, who is still using her legal knowledge for good. She is now serving as first deputy attorney general of New York State, working under the progressive Letitia James. Her family thrives. Since charities are under the remit of the attorney general, Jennifer no longer serves on the board of ILLAHIE, even though it is Canadian. Ben still lives up the coast with his second wife, both involved in good works through their church activities. His children are grown now and each is starting their journey, facing a world that is radically different from the one I faced at their age.

My journey is not over yet, although I'm now in my eighties and we are spending much of the time in Lund, BC, a little fishing village 150 kilometers up the coast from Vancouver. Even though there is so much talent in Canada and in BC, it is still extraordinarily difficult to raise even small amounts of capital for start-up biotech companies. And it's equally difficult to establish even a beachhead for mature companies in our country. Successful companies that start in Canada are, more often than not, bought by larger US industries. This happens in all sectors, not just high tech and biotech. It is the nature of businesses in general to grow by any means, including by absorbing other smaller businesses. Canadian companies often end up being those smaller entities. So perhaps the dream of creating a sustainable and strong biotech industry in Canada is unrealistic. Be that as it may, I will go on trying as long as I can.

Acknowledgments

My special thanks go to Ellen Godfrey, my friend and rigorous editor who patiently taught me how to write a memoir. I also want to acknowledge the thoughtful and generous comments made by Dr. Molly Shoichet in her foreword. Thank you also to my many friends and colleagues from QLT who gave me guidance and encouragement while I was writing. These people include Karole Sutherland, Linda Lupini, David Main, and Paul Hastings, among many others not mentioned. And a special thank you to Ian Gill, Dr. Bill Leiss, and Mark Jones for helping and guiding in navigating the publishing industry.

About the Author

Julia Levy earned her PhD at the age of twenty-three from University College, University of London. She taught microbiology and immunology in the department of microbiology and established a well-funded research lab at the University of British Columbia. In 1980 she joined with four colleagues to form the biotechnology company QLT Inc. as a university spin-off. Discoveries made by her and fellow scientists at the university in the field of photodynamic therapy led to the first approved medical treatment for age-related macular degeneration, the commonest cause of severe vision loss in the elderly. She served first as Chief Scientific Officer and later as Chief Executive Officer of QLT. During her tenure as CEO the company obtained regulatory approval for its macular degeneration treatment that resulted in the financial success of QLT She is an Officer of the Order of Canada and has received a number of honorary degrees. A leader in the fields of science, education, and business, Dr. Levy has served on the boards of a number of companies as well as charitable organizations.

Index